How Cities Learn

RGS-IBG Book Series

For further information about the series and a full list of published and forthcoming titles please visit www.rgsbookseries.com

Published

How Cities Learn: Tracing Bus Rapid Transit in South Africa
Astrid Wood

Defensible Space on the Move: Mobilisation in English Housing Policy and Practice
Loretta Lees and Elanor Warwick

Geomorphology and the Carbon Cycle
Martin Evans

The Unsettling Outdoors: Environmental Estrangement in Everyday Life
Russell Hitchings

Respatialising Finance: Power, Politics and Offshore Renminbi Market Making in London
Sarah Hall

Bodies, Affects, Politics: The Clash of Bodily Regimes
Steve Pile

Home SOS: Gender, Violence, and Survival in Crisis Ordinary Cambodia
Katherine Brickell

Geographies of Anticolonialism: Political Networks Across and Beyond South India, c. 1900-1930
Andrew Davies

Geopolitics and the Event: Rethinking Britain's Iraq War through Art
Alan Ingram

On Shifting Foundations: State Rescaling, Policy Experimentation And Economic Restructuring In Post-1949 China
Kean Fan Lim

Global Asian City: Migration, Desire and the Politics of Encounter in 21st Century Seoul
Francis L. Collins

Transnational Geographies Of The Heart: Intimate Subjectivities In A Globalizing City
Katie Walsh

Cryptic Concrete: A Subterranean Journey Into Cold War Germany
Ian Klinke

Work-Life Advantage: Sustaining Regional Learning and Innovation
Al James

Pathological Lives: Disease, Space and Biopolitics
Steve Hinchliffe, Nick Bingham, John Allen and Simon Carter

Smoking Geographies: Space, Place and Tobacco
Ross Barnett, Graham Moon, Jamie Pearce, Lee Thompson and Liz Twigg

Rehearsing the State: The Political Practices of the Tibetan Government-in-Exile
Fiona McConnell

Nothing Personal? Geographies of Governing and Activism in the British Asylum System
Nick Gill

Articulations of Capital: Global Production Networks and Regional Transformations
John Pickles and Adrian Smith, with Robert Begg, Milan Buček, Poli Roukova and Rudolf Pástor

Metropolitan Preoccupations: The Spatial Politics of Squatting in Berlin
Alexander Vasudevan

Everyday Peace? Politics, Citizenship and Muslim Lives in India
Philippa Williams

Assembling Export Markets: The Making and Unmaking of Global Food Connections in West Africa
Stefan Ouma

Africa's Information Revolution: Technical Regimes and Production Networks in South Africa and Tanzania
James T. Murphy and Pádraig Carmody

Origination: The Geographies of Brands and Branding
Andy Pike

In the Nature of Landscape: Cultural Geography on the Norfolk Broads
David Matless

Geopolitics and Expertise: Knowledge and Authority in European Diplomacy
Merje Kuus

Everyday Moral Economies: Food, Politics and Scale in Cuba
Marisa Wilson

Material Politics: Disputes Along the Pipeline
Andrew Barry

Fashioning Globalisation: New Zealand Design, Working Women and the Cultural Economy
Maureen Molloy and Wendy Larner

Working Lives - Gender, Migration and Employment in Britain, 1945-2007
Linda McDowell

Dunes: Dynamics, Morphology and Geological History
Andrew Warren

Spatial Politics: Essays for Doreen Massey
Edited by David Featherstone and Joe Painter

The Improvised State: Sovereignty, Performance and Agency in Dayton Bosnia
Alex Jeffrey

Learning the City: Knowledge and Translocal Assemblage
Colin McFarlane

Globalizing Responsibility: The Political Rationalities of Ethical Consumption
Clive Barnett, Paul Cloke, Nick Clarke & Alice Malpass

Domesticating Neo-Liberalism: Spaces of Economic Practice and Social Reproduction in Post-Socialist Cities
Alison Stenning, Adrian Smith, Alena Rochovská and Dariusz Świątek

Swept Up Lives? Re-envisioning the Homeless City
Paul Cloke, Jon May and Sarah Johnsen

Aerial Life: Spaces, Mobilities, Affects
Peter Adey

Millionaire Migrants: Trans-Pacific Life Lines
David Ley

State, Science and the Skies: Governmentalities of the British Atmosphere
Mark Whitehead

Complex Locations: Women's geographical work in the UK 1850–1970
Avril Maddrell

Value Chain Struggles: Institutions and Governance in the Plantation Districts of South India
Jeff Neilson and Bill Pritchard

Queer Visibilities: Space, Identity and Interaction in Cape Town
Andrew Tucker

Arsenic Pollution: A Global Synthesis
Peter Ravenscroft, Hugh Brammer and Keith Richards

Resistance, Space and Political Identities: The Making of Counter-Global Networks
David Featherstone

Mental Health and Social Space: Towards Inclusionary Geographies?
Hester Parr

Climate and Society in Colonial Mexico: A Study in Vulnerability
Georgina H. Endfield

Geochemical Sediments and Landscapes
Edited by David J. Nash and Sue J. McLaren

Driving Spaces: A Cultural-Historical Geography of England's M1 Motorway
Peter Merriman

Badlands of the Republic: Space, Politics and Urban Policy
Mustafa Dikeç

Geomorphology of Upland Peat: Erosion, Form and Landscape Change
Martin Evans and Jeff Warburton

Spaces of Colonialism: Delhi's Urban Governmentalities
Stephen Legg

People/States/Territories
Rhys Jones

Publics and the City
Kurt Iveson

After the Three Italies: Wealth, Inequality and Industrial Change
Mick Dunford and Lidia Greco

Putting Workfare in Place
Peter Sunley, Ron Martin and Corinne Nativel

Domicile and Diaspora
Alison Blunt

Geographies and Moralities
Edited by Roger Lee and David M. Smith

Military Geographies
Rachel Woodward

A New Deal for Transport?
Edited by Iain Docherty and Jon Shaw

Geographies of British Modernity
Edited by David Gilbert, David Matless and Brian Short

Lost Geographies of Power
John Allen

Globalizing South China
Carolyn L. Cartier

Geomorphological Processes and Landscape Change: Britain in the Last 1000 Years
Edited by David L. Higgitt and E. Mark Lee

How Cities Learn

Tracing Bus Rapid Transit in South Africa

Astrid Wood

WILEY

This edition first published 2022
© 2022 Royal Geographical Society (with the Institute of British Geographers)
This Work is a co-publication between The Royal Geographical Society (with the Institute of British Geographers) and John Wiley & Sons Ltd.

All rights reserved. No part of this publication may be reproduced, stored in a retrieval system, or transmitted, in any form or by any means, electronic, mechanical, photocopying, recording or otherwise, except as permitted by law. Advice on how to obtain permission to reuse material from this title is available at http://www.wiley.com/go/permissions.

The right of Astrid Wood to be identified as the author of this work has been asserted in accordance with law.

Registered Office(s)
John Wiley & Sons, Inc., 111 River Street, Hoboken, NJ 07030, USA
John Wiley & Sons Ltd, The Atrium, Southern Gate, Chichester, West Sussex, PO19 8SQ, UK

Editorial Office
9600 Garsington Road, Oxford, OX4 2DQ, UK

For details of our global editorial offices, customer services, and more information about Wiley products visit us at www.wiley.com.

Wiley also publishes its books in a variety of electronic formats and by print-on-demand. Some content that appears in standard print versions of this book may not be available in other formats.

Limit of Liability/Disclaimer of Warranty
The contents of this work are intended to further general scientific research, understanding, and discussion only and are not intended and should not be relied upon as recommending or promoting scientific method, diagnosis, or treatment by physicians for any particular patient. In view of ongoing research, equipment modifications, changes in governmental regulations, and the constant flow of information relating to the use of medicines, equipment, and devices, the reader is urged to review and evaluate the information provided in the package insert or instructions for each medicine, equipment, or device for, among other things, any changes in the instructions or indication of usage and for added warnings and precautions. While the publisher and authors have used their best efforts in preparing this work, they make no representations or warranties with respect to the accuracy or completeness of the contents of this work and specifically disclaim all warranties, including without limitation any implied warranties of merchantability or fitness for a particular purpose. No warranty may be created or extended by sales representatives, written sales materials or promotional statements for this work. The fact that an organization, website, or product is referred to in this work as a citation and/or potential source of further information does not mean that the publisher and authors endorse the information or services the organization, website, or product may provide or recommendations it may make. This work is sold with the understanding that the publisher is not engaged in rendering professional services. The advice and strategies contained herein may not be suitable for your situation. You should consult with a specialist where appropriate. Further, readers should be aware that websites listed in this work may have changed or disappeared between when this work was written and when it is read. Neither the publisher nor authors shall be liable for any loss of profit or any other commercial damages, including but not limited to special, incidental, consequential, or other damages.

Library of Congress Cataloging-in-Publication Data
Names: Wood, Astrid, author. | Royal Geographical Society (Great Britain),
 issuing body. | Institute of British Geographers, issuing body.
Title: How cities learn : tracing bus rapid transit in South Africa /
 Astrid Wood.
Description: Hoboken, NJ : John Wiley & Sons, Inc., 2022. | Includes
 bibliographical references and index.
Identifiers: LCCN 2021044851 (print) | LCCN 2021044852 (ebook) | ISBN
 9781119794271 (hardback) | ISBN 9781119794288 (paperback) | ISBN
 9781119794318 (pdf) | ISBN 9781119794301 (epub) | ISBN 9781119794295
 (ebook)
Subjects: LCSH: Bus rapid transit--South Africa. | Transportation--Social
 aspects--South Africa.
Classification: LCC HE5704.4.A6 W66 2022 (print) | LCC HE5704.4.A6
 (ebook) | DDC 388.4/13220968--dc23/eng/20211130
LC record available at https://lccn.loc.gov/2021044851
LC ebook record available at https://lccn.loc.gov/2021044852

Cover image: © Astrid Wood
Cover design by Wiley

Set in 10/12pt Plantin Std by Integra Software Services Pvt. Ltd, Pondicherry, India
Printed and bound by CPI Group (UK) Ltd, Croydon, CR0 4YY

C115759_150322

Contents

List of Figures viii
List of Abbreviations x
Series Editors' Preface xii
Acknowledgements xiii

1. **Introduction** 1
   BRT Arrives in South Africa 1
   Understanding the South African City 2
   Transport Geography, Policy Mobilities and Learning in and from the South 5
   Using Policy Mobilities as a Methodology 9
   Structure of the Book 13

2. **Geographies of Knowledge** 16
   Building an Analytic for Tracing 16
   Tracing through Policy Models 18
   Tracing through Actors and Associations 20
   Tracing through Cities 23
   Tracing through Temporalities 25

3. **Translating BRT to South Africa** 27
   Introduction 27
   The Geography of BRT 28
   Forming the Bogotá Model of BRT 31
   Introducing BRT in South African Cities 35
      Johannesburg's Rea Vaya 36
      Cape Town's MyCiTi 39
      Tshwane's A Re Yeng 41
      Rustenburg's Yarona 44
      Nelson Mandela Bay's Libhongolethu 45

eThekwini's Go Durban!	46
A South African Interpretation of BRT	48
About the Station Platform	51
About the Bus	52
About the Bus Lane	53
About the Route	55
BRT and Taxi Transformation	58
The South African Taxi Industry	59
State Intervention in Transportation	61
Negotiating with Taxi Operators	65
Conclusion	68
4. Actors and Associations Circulating BRT	**70**
Introduction	70
An Analytic for Studying Policy Actors	71
Redefining the Role of Policy Actors	74
Policy Mobilizers of BRT Circulation	75
Intermediaries of BRT Circulation	78
Local Pioneers of BRT Circulation	81
Learning through Networks	85
Networks of Internationals	86
Networks of South Africans	88
Power Dynamics of Networks	94
Conclusion	96
5. The Local Politics of BRT	**97**
Introduction	97
The International Context of BRT Circulation	98
Learning from South America	99
Learning from Africa	102
Learning from India	105
Learning from the North	106
The National Context of BRT Circulation	107
Political Interactions between South African Localities	108
Technical Exchanges between South Africa Localities	111
The Municipal Context of BRT Circulation	114
Conclusion	117
6. Repetitive Processes of BRT Adoption	**119**
Introduction	119
Tracing Transportation Innovation in South Africa	120
Planting the Seeds of BRT in South Africa	124
Gradual Processes of Learning	127

	Repetitive Processes of Circulation	128
	Delayed Processes of Adoption	130
	Transportation Innovations Not Adopted	133
	Conclusion	138
7.	**Conclusion**	**140**
	Introduction	140
	Reflecting on How Cities Learn	141
	Reflecting on BRT in South Africa	145

Appendix A: Interview Schedule — 147
Appendix B: Features of BRT systems in South Africa — 154
References — 166
Index — 185

List of Figures

Figure 3.1	Number of BRT systems opening annually	32
Figure 3.2	Map of BRT in South Africa	35
Figure 3.3	BRT adoption and implementation in South Africa	36
Figure 3.4	Fashion Square Rea Vaya station, Johannesburg	38
Figure 3.5	Lagoon Beach MyCiTi station, Cape Town	40
Figure 3.6	Map of A Re Yeng, Tshwane	42
Figure 3.7	Hatfield A Re Yeng station, Tshwane	43
Figure 3.8	Yarona station platform, Rustenburg	45
Figure 3.9	Features of BRT Systems in Cape Town and Johannesburg	49
Figure 3.10	Rea Vaya high-floor station, Johannesburg	52
Figure 3.11	Rea Vaya bus, Johannesburg	53
Figure 3.12	Rea Vaya bus lane, Johannesburg	54
Figure 3.13	MyCiTi bus lane, Cape Town	55
Figure 3.14	Map of the MyCiti, Cape Town	56
Figure 3.15	Map of Rea Vaya, Johannesburg	57
Figure 3.16	Modal split in South African cities	61
Figure 3.17	Public Transport Infrastructure and Systems Grant allocation	63
Figure 3.18	BRT in political cartoons	66
Figure 4.1	Types of policy actors	73
Figure 4.2	BRT policy actors	75
Figure 4.3	'Who told you about BRT?'	86
Figure 4.4	Details of South African municipal BRT-related study tours	90
Figure 4.5	Percent of respondents who went on a study tour to Bogotá	91
Figure 5.1	Shekilango BRT Station, Dar es Salaam	104

Figure 5.2	Learning process across South African cities	114
Figure 6.1	Horse-drawn tram in Johannesburg	121
Figure 6.2	Electric trams in Johannesburg	122
Figure 6.3	Knowledge of BRT adoption in South Africa	124
Figure 6.4	Exclusive curb lane on city streets	127

List of Abbreviations

ACET	African Centre of Excellence for Studies of Public and Non-motorized Transport/Centre for Transport Studies
ALC-BRT	Across Latitudes and Cultures – Bus Rapid Transit
ANC	African National Congress
BEN	Bicycle Empowerment Network
BIA/BID	business improvement area/business improvement district
BRT	bus rapid transit
CSIR	Council for Scientific and Industrial Research
DA	Democratic Alliance
ETA	eThekwini Transport Authority
FWC	football world cup
GCRO	Gauteng City Region Observatory
GIZ	German Society for International Cooperation
HSRC	Human Sciences Research Council
IDP	integrated development plans
IIEC	International Institute for Energy Conservation
IPPUC	Curitiba Research and Urban Planning Institute
IRT	integrated rapid transit
ITDP	Institute for Transportation and Development Policy
ITP	integrated transport plans
ITS	intelligent transportation systems
JDA	Johannesburg Development Authority
JIKE	Johannesburg Innovation and Knowledge Exchange
JRA	Johannesburg Roads Authority
LAMATA	Lagos Metropolitan Area Transport Authority
MIIF	Municipal Infrastructure Investment Framework
MILE	Municipal Institute of Learning

MMC	Member of Mayoral Committee
NLTA	National Land Transport Act (2009)
NLTTA	National Land Transport Transition Act (2000)
PRASA	Passenger Rail Agency of South Africa
PRT	personal rapid transit
PTISG	Public Transport Infrastructure and Systems Grant
PUTCO	Public Utility Transport Corporation
SABOA	South African Bus Operators Association
SACN	South African Cities Network
SALGA	South African Local Government Association
SANRAL	South African Roads Association Ltd.
SATC	Southern African Transport Conference
SPTN	Strategic Public Transport Network
TRP	Taxi Recapitalization Program
UCLG	United Cities and Local Governments
UCT	University of Cape Town
UK	United Kingdom

Series Editors' Preface

The RGS-IBG Book Series only publishes work of the highest international standing. Its emphasis is on distinctive new developments in human and physical geography, although it is also open to contributions from cognate disciplines whose interests overlap with those of geographers. The series places strong emphasis on theoretically informed and empirically strong texts. Reflecting the vibrant and diverse theoretical and empirical agendas that characterize the contemporary discipline, contributions are expected to inform, challenge and stimulate the reader. Overall, the RGS-IBG Book Series seeks to promote scholarly publications that leave an intellectual mark and change the way readers think about particular issues, methods or theories.

For details on how to submit a proposal please visit: www.rgsbookseries.com

Ruth Craggs, *King's College London, UK*
Chih Yuan Woon, *National University of Singapore*
RGS-IBG Book Series Editors

David Featherstone
University of Glasgow, UK
RGS-IBG Book Series Editor (2015–2019)

Acknowledgements

This book is the outcome of more than a decade of research and it would be impossible to thank all those whose support has buoyed it.

To my interviewees in South Africa and those who gave me so many hours of their time, I hope this work provides support to continue improving urban life. This book captures and shares the story of BRT in South Africa, and provides a vital record of post-apartheid transformation. Many of the key figures named in this book have since left government, retired or otherwise moved on to new positions, taking their institutional memory with them. They risked their careers and their lives for a more equitable South Africa. Thank you for your bravery, dedication and candor.

I am indebted to my mentors at Newcastle University and UCL as well as Cardiff University, Royal Holloway University of London and London School of Economics for their feedback and encouragement. I am eternally grateful to Jenny Robinson and Andrew Harris whose insightful comments provided the roadmap throughout my learning process.

I am thankful to my friends and colleagues in Newcastle, London, Cape Town, Johannesburg, and around the world who offered feedback on journal articles, book chapters, conference presentations and grant proposals, as well as direction for how to navigate the academic netherworld.

A special thanks to the series editors Ruth Craggs, David Featherstone and Chih Yuan Woon for their careful engagement with the manuscript.

I am lucky to have an encouraging husband whose love is my firmest support, children who brighten even the cloudiest of days, and a family that keeps me grounded throughout the many travels we take as academics and individuals.

This book captures and shares the story of BRT in South Africa, and provides a vital record of post-apartheid transformation. Many of the key figures named in this book have since left government, retired or otherwise moved on to new positions, taking their institutional memory with them. They risked their careers and their lives for a more equitable South Africa. Thank you for your bravery, dedication and candor.

I dedicate this work to my late father whose pursuit of social justice imbued me with a similar sense. His anti-apartheid activism, along with so many others, helped make the country of my birth a better place.

This book is for all those who believe.

Chapter One
Introduction

BRT Arrives in South Africa

From Curitiba and Bogotá to Ahmedabad and Beijing, bus rapid transit (BRT) has promised to be a quick, cost-effective and efficient method of urban transportation that combines the speed and quality of rail transportation with the flexibility of a bus system. BRT is a rubber-tired mode of urban public transportation that combines buses, busways and stations with intelligent transportation systems, operational and financial plans, integrated ticketing, and a branded identity. It has been a dominant feature of urban planning for decades in cities as diverse as Bogotá, Curitiba, Guangzhou, Lima, Los Angeles, Mumbai, and New York, among others. Whereas previous studies have considered the characteristics of BRT (Deng and Nelson 2011; Jarzab et al. 2002; Levinson et al. 2003) or its impact on transportation planning (Ferbrache 2019; Paget-Seekins and Munoz 2016), this book is the first attempt to understand the global proliferation of BRT.

Much of its current popularity is due to the vehement promotion undertaken by Enrique Penalosa, Bogotá's Mayor from 1998 to 2001 and again from 2016 to 2019, and his ties with the Institute for Transportation and Development Policy (ITDP) (Wood 2014b, 2019b). More than two decades since Bogotá's Transmilenio opened to global acclaim, BRT has become one of the most prominent policy solutions of the 21st century. Around the world, Transmilenio-style systems are commended by BRT advocates for improving mobility, by reducing travel time and improving comfort and reliability; and its transformation into best practice is often attributed to its affordability, brief implementation phase and gener-

How Cities Learn: Tracing Bus Rapid Transit in South Africa, First Edition. Astrid Wood.
© 2022 Royal Geographical Society (with the Institute of British Geographers). Published 2022 by John Wiley & Sons Ltd.

ous political payoffs. It is presented as a best practice appropriate within a variety of geographical and socio-political settings, and able to tackle problems related to economic exclusion and inequality, urban sprawl and sustainability, and transportation inaccessibility.

The Bogotá model of BRT first arrived in South Africa in July 2006 at a special session of the Southern African Transport Conference (SATC), the largest transportation convention in the region and a critical platform for dialogue on issues ranging from finance to public transportation. Lloyd Wright, a global expert on BRT, was invited by the National Department of Transport to host a day-long workshop on the principles, attributes and engineering specifications of BRT. This learning was reinforced in August 2006 when Lloyd Wright visited politicians and transportation planners in Cape Town, eThekwini, Johannesburg and Tshwane to present the attributes of BRT. Interested cities then took a select group of politicians, planners, operators and consultants to Bogotá to see how BRT operates and meet with transportation operators. Policymakers returned from these study tours eager to introduce BRT locally.

Since 2006, BRT has been adopted in six South African cities to improve transportation services, especially for the urban poor. Cape Town, eThekwini, Johannesburg, Nelson Mandela Bay, Rustenburg, and Tshwane are currently in various stages of planning and implementation: in August 2009, just three years after learning of the Bogotá model of BRT, Rea Vaya Phase 1A opened in Johannesburg as the first full-feature BRT system in an African context; in May 2011, Cape Town's MyCiTi Phase 1A became operational; in May 2012, eThekwini Council approved plans to proceed with the first three lines of Go Durban!; and in July 2012, the cascade continued with Rustenburg and Tshwane beginning construction on Yarona and A Re Yeng. Not all cities have had a simple, straightforward experience, however: since 2008, Nelson Mandela Bay's attempt to introduce BRT has been stalled by municipal politics and poor planning, and in spite of considerable efforts, the project remains in a state of postponement.

While the South African systems are unmistakably modeled after the achievements of those in Bogotá, the process through which South African officials learned of, and implemented BRT, remains unexplored. In mapping the learning process, this book considers how and why city leaders adopt circulated best practice.

Understanding the South African City

The adoption of BRT in South Africa reflects the historical spatial planning of apartheid (Christopher 1995; Parnell 1997; Parnell and Mabin 1995; Robinson 1996, 1997) and the challenges facing post-apartheid policies to remedy these dysfunctional schemes (Haferburg and Huchzermeyer 2014; Harrison et al. 2008, 2014; Parnell and Pieterse 2014). South African cities were shaped primarily by policies of strict racial segregation but also rigorous separation of economic and residential zones, which denied Black residents full access to the city and its economic base (Davies 1981; Home 1990; Lemon 1991; Western 1985); and had the secondary objective of increasing travel times considerably for non-White residents (Pirie 2013, 2014). Apartheid settlement strategies located townships on the periphery of cities and heavily subsidized public transportation to enable workers to travel long distances at low fares (Beall et al. 2002; Turok and Watson 2001). Under the old regime, only those with passes were permitted to travel between the townships and the city, and thus movement was generally only permitted on weekdays between home and work. Because of the inheritance of these restrictions, to this day there is effectively no pattern of non-work travel between the suburban areas and the city center. The introduction of BRT is an attempt to unsettle these socio-spatial settlement patterns.

Today, South African cities are characterized by contrasts and dualisms: high-rise residential towers turned slums; Victorian houses surrounded by privatized greenery; endless stretches of banal suburban development punctured by low-cost government-sponsored housing; European cafés and upmarket shops with hawkers selling homemade wares and promising to guard the luxury cars. The one commonality across the fragmented post-apartheid landscape is the proliferation of the automobile – its presence dominates the physical landscape of the city as well as the cultural milieu. Obviously, the South African city is not unique in this feature, but the degree to which apartheid's forced segregation stretched the city amplifies this condition. Although this understanding of the spatial character of the South African city as uneven is generally applied ubiquitously, there are profound differences across South African cities reflecting their distinctive topography and resulting settlement patterns, as well as their sociocultural composition, economic vitality and historic planning and contemporary governance. My assessment of the spatial form and associated mobility dynamics sheds light on the complex and challenging advancement of inclusive South African cities.

South African history is riddled with transportation experiments: horse-drawn streetcars were introduced in the 1890s, electric trams operated until the Second World War and trolleybuses ran in the high-apartheid period until Johannesburg, the last city to do so, terminated services in 1986. There are a number of detailed empirical accounts of the trams and trolleybus systems in Cape Town (Gill 1961; Joyce 1981), eThekwini (Jackson 2003), Johannesburg (Sey 2012; Spit and Patton 1976) and Nelson Mandela Bay (Shields 1979), as well as analyses of the development of the road system (e.g. Rosen 1962 in Johannesburg) and the emergence of the minibus taxi industry (Khosa 1991, 1995; McCaul 1990). Because South African city form and function makes it difficult to support a sunken subway – Johannesburg is built atop a maze of underground gold mining shafts and Cape Town rests largely on marsh and infill – transportation officials and engineers have struggled to modernize the commuter rail and municipal bus services. The commuter rail network is poorly maintained and its fixed lines prove inadequate in the expanding metropolises. Bus systems are similarly struggling to service the low-density urban form. The modernist aspiration of car ownership and its associations with independence and wealth is reinforced in practical terms through the dispersed city form, which separates people from economic and social opportunities.

For the most part, the urban populace relies on a politically powerful and largely under-regulated fleet of overcrowded, poorly maintained minibus taxis that operate irregular services. The minibus taxi industry has captured the majority of market share against subsidized modes, carrying about 60 percent of trips, nationally. The industry emerged in the 1980s in reaction to the failures of government to supply adequate bus and train services to the townships (Khosa 1991, 1995; McCaul 1990). In the sprawling landscape of contemporary urban South Africa, the minibus taxi is generally preferred to government-sponsored bus and rail services because it is considered more convenient in terms of routing and frequency (Clark and Crous 2002). While there are certainly arguments in support of the minibus taxi industry with proponents describing it as a self-made, Black entrepreneurial venture, in general commuters are dissatisfied with the slow, capricious quality of the informal services (Salazar Ferro et al. 2013). The South African policymakers I interviewed described an almost doomsday scenario filled with uncertainty, labeling it a "commuter crisis" akin to the global financial crisis and calling for fundamental reform to the transportation network.[1]

As a result of these features, transportation planning has been understood as central to the transformation of South African cities. Transportation has historically been used to divide the spatial layout of cities. Planned roads have been used to separate planning typologies in both planned and unplanned settlements, and transportation systems have been used to control who can access the city and how they move. In South Africa, planners have been especially focused on building modernist highways to accommodate the White elite who could afford to drive. The fact that transportation is experienced by a range of people across incomes and experiences, means that it also serves as an arena for social mixing and these interactions have unbridled opportunities for change. Policymakers in South Africa have been attuned to these openings, and efforts to remedy the inequality of transportation have been at the forefront of urban planning and policymaking since 1994. Transportation has also been, and continues to be, a site of resistance. The success of the 1957 Alexandra bus boycotts in Johannesburg was a pivotal moment in the anti-apartheid movement; and in the post-apartheid era, transportation continues to be a point of contention in service delivery protests.

This book aims to explain how South African policymakers are trying to improve urban transportation, specifically addressing the process by which best practice is drawn from elsewhere to inform local planning and policy change. In South Africa, BRT is seen as the solution that simultaneously provides transportation users with an affordable, reliable and safe transportation system, taxi operators with formalized and stable employment, and buses and rail operators with viable routes. Its larger purpose is to address the severe historical spatial divide along racial lines and post-apartheid splintered urbanism. The operational systems conjure up images of equality and dignity for all South Africans, moving freely and efficiently through urban space regardless of skin color or income, in a city free from the grip of informality, and instead managed by an efficient and capable municipal government. In this post-apartheid moment, transportation may be South Africa's best tool through which to bridge the divided city.

Transport Geography, Policy Mobilities and Learning in and from the South

City learning is hardly a new practice. Herodotus described information exchanges as early as 500 BCE; in the second century, Palmyra adapted Roman concepts of urbanity; and in the 1700s, Peter the Great employed Dutch architectural models in St. Petersburg (Healey 2013). In the early 20th century, cities shared their experiments with electricity, gas, sewerage and water services (Dogliani 2002; Gaspari 2002; Kozinska-Witt 2002; Saunier 2002; Vion 2002). These exchanges became a "precious resource" to subvert or strengthen local policy decisions (Saunier 2002: 519). These "transboundary connections" (Saunier 2002: 510) were often a method of "intergovernmental diplomacy" (Saunier 2002: 509), with scholars suggesting that these collaborations advanced urban development (Healey and Upton 2010; Saunier 2002; Saunier and Ewen 2008; Sutcliffe, 1981).

In the contemporary era, learning has become a regular and routine aspect of policy creation. Policymakers seek innovative lessons and models from elsewhere, assuming that "viewing a familiar problem in an unfamiliar setting expands ideas of what is possible, and can inspire fresh thinking about what to do at home" (Rose 1993: 30). Rose (1993) suggests that positive lessons provide insight for local policymakers, and negative lessons help them avoid others' mistakes. And Bennett (1991) confirms that when cities are confronted with local challenges, there is a natural tendency to look elsewhere for innovation. Learning, however, is deeply entangled with the politics of people and place, and it should be understood in terms of the cultural, economic, historical, and political connections and disconnections through which knowledge moves. Cities and their policymakers compete for prestige, and practices of exchange are rarely just about rationalist transfers of knowledge. Mobilized policy is often seen as "politically neutral truths", but beneath "this superficial impartiality" these lessons can serve as "political weapons" (Robertson 1991). Indeed, acts of knowledge sharing initiate particular policies and certain cities into conversations, while pushing others apart.

The South African city's propensity to apply foreign planning models is rooted in its history, where urban design and transportation innovations were imported from the colonial metropole. Colonialism created an atmosphere conducive to the temptation of imitation, in which the local environment lacked "genuine ties with the world surrounding them" (Mbembe 2004: 375), and instead linked itself to classical aspects of European cosmopolitanism. This is evident in architectural form: various technical and political interventions including "prefabricated iron-fronted shop buildings, barrel-vaulted arcades with prismatic glass skylights, cast-iron gas lamps, electric lighting, telephone wires…" (Chipkin 1993: 22), as well as horse-drawn trams and railroads. This "overseas cultural traffic" (Mbembe and Nuttall 2004: 362) flowed through colonial relationships with London, Paris and New York. During the emergence of professionalized planning, universities trained colonial planners in modernist techniques, later applied across the colonial world. In South Africa, town planners studied in London after the Second World War and, through these exchanges, developed what became classic apartheid planning mechanisms (Wood 2019a).

Today, the South African urban landscape is reflective of a global convergence of policy knowledge, and several ostensibly South African policies are also evidenced elsewhere: approaches to growth management, informal settlement upgrading, sustainability, and even securitization and gated communities (Bénit-Gbaffou et al. 2012; Morange et al. 2012) migrated from North America, Brazil and Europe, and were adopted in South African cities because of their success elsewhere; city improvement districts (elsewhere, business improvements districts) are located in precincts across Cape Town and Johannesburg, as well as in Amsterdam, London and New York (Didier et al. 2012; Morange et al. 2012; Peyroux et al. 2012) and city development strategies, in particular Johannesburg's 2040 Growth and Development Strategy, are considered best practice and duplicated globally (Robinson 2011). While the regularity by which South African cities learn of and implement policies from elsewhere is evident, the process of, and rationale for, learning and adoption demands further theoretical unpacking.

In South Africa, a desire to copy from urban experiences elsewhere is reinforced today through the language of south–south exchange, which suggests that localities across the global south may find better solutions by looking to one another rather than to their colonial metropole. Southern cities are presumed to share commonality with their postcolonial neighbors, and by exchanging their experiences and experiments, it is thought that they might develop more imaginative and effective solutions to remedy their urban challenges. This political argument has fueled the circulation of supposedly southern-generated best practices like BRT. Questions remain, however, regarding the extent to which this learning impacts development in southern cities, either by reinforcing former colonial ties through contemporary practices of exchange or by shattering those dependencies, instead generating southern solutions to southern problems. Elsewhere I have argued that south–south learning may lead cities toward more effective policy solutions, but efforts need to be made to ensure that learning is not merely political window-dressing (Wood 2015b). This is especially important in providing substance to support localities in their determination to reject apparent best practices. Evidence from this book supports contemporary academic and practical efforts at decentering epistemic knowledge, by encouraging (South African) cities to draw from a wider array of examples from within South Africa and across the postcolonial world.

A focus on BRT adoption provides an opportunity to reinterpret both the historical and contemporary South African city as a site of "mimicry" and "mimesis" (Mbembe 2004). Mbembe (2008: 7) suggests "if there is ever an African form of metropolitan modernity, then Johannesburg will have been its classical location"; and Robinson (2003: 260) concludes, "Johannesburg is an antidote to [a] divisive tradition in urban studies and a practical example of how cities can be imagined outside of the global/developmentalist division". Mimicry, however, does not occur simply because reforms from elsewhere are better, but rather because the very action of copying may accelerate local policymaking. Nevertheless, Mbembe argues that even cities "born out of mimicry are capable of mimesis", by establishing "similarities with something else while at the same time inventing something original" (2004: 376). This helps explain how South African cities learn of a policy or practice from elsewhere, transfer it across boundaries, and localize it to suit the South African city.

How Cities Learn contributes to efforts to transform transport geography into a more inclusive and global endeavor, by examining the production and distribution of transportation knowledge in the global south. In a related project, we argue against the continued dominance of northern transportation models and best practices, and instead highlight locally derived experiments in both the global north and south (Wood et al. 2020). This means not only featuring the achievements of cities that are "off the map" (Robinson 2006), but also de-centering the locations in which best practice is solidified and sent forth. A decolonial approach to transport geography challenges its technocratic objectivism and mathematical modeling, which limit the promotion of southern experiments. Transportation scholars have begun to engage with these strategies – for instance, a special issue featuring urban mobilities in the global south (Priya Uteng and Lucas 2018) – but much more needs to be said of transport geography from, of and by the global south.

This book is grounded within South African urban transportation research. It draws on calculations of shifting mobility patterns by Roger Behrens in Cape Town (2013, 2014, 2015), Jackie Walters in Johannesburg (2009, 2013) and Christoph Venter in Tshwane (2013), as well as Gordon Pirie's (2013, 2014) studies of transportation practices, policies and perspectives. Recent studies of BRT user experience (Behrens and Wilkinson 2003; Maunganidze and Del Mistro 2012; Schalekamp and Behrens 2008), government efforts to reform the taxi industry (Schalekamp and Behrens 2013; Schalekamp et al. 2010), cycling

(Jennings 2015) and transit-oriented development (Bickford and Behrens 2015), also provide the bedrock for this analysis. Additionally, it overlaps BRT adoption with broader considerations for rapid urbanization across the continent, infrastructure development and its potential for poverty alleviation, climate change and resilience, especially in disenfranchised communities, and the juxtaposition between urban economic vitality and social justice.

This book employs "policy mobilities" (Peck and Theodore 2010a; McCann and Ward 2011) to contribute to transport geography by examining how policymakers address issues of mobility and immobility, and how these decisions are made in reference to similar practices taking place elsewhere. It explains the process by which BRT has been embraced, encompassed and even at times excluded by local policy actors, their interactions with global advocates and inter-referencing across space and time. In so doing, it attends to the ways in which transportation solutions are engineered in relation to socio-political spaces, specifically interrogating the process through which certain transportation innovations are deemed best practice.

How Cities Learn makes four key contributions to the policy mobilities literature by examining the process from one location: South Africa. First, it highlights the ways in which particular models of best practice travel, not autonomously by virtue of their own universalist qualities, but rather via a complex political economy, both internationally and domestically.

Second, the book exposes the pivotal role of often overlooked local actors in the mobility and adoption of best practice. While it might seem as if international intermediaries are the primary actors involved in BRT replication, *How Cities Learn* demonstrates the power of the local. Although strong personalities were at times persuasive in pushing BRT from the outside, their dominance was balanced by the bureaucratic structures that endow local implementers with the power and responsibility for initiating a new planning project. Notably, this approach highlights the factors motivating policy actors and the wider political relationships that facilitated the reception of BRT. This in turn expands our understanding of the varied direction, speed and influence of global and local influences shaping the contemporary city.

Third, the book unpacks the exchanges between actors in South African and other southern cities and, in so doing, exposes a politicized process that preferences certain sites over others, inculcates a particular understanding of best practice and otherwise mediates policy flows. The politics of south-south cooperation was used to support exchanges with South American cities while overlooking opportunities to learn from African and Indian cities. These glimpses into the process of BRT adoption in South Africa help us understand the process by which policy and policy actors connect and disconnect topographically through their physical travels, as well as topologically through relational comparisons made in city rankings and league tables.

Fourth, the book uncovers the temporalities of policy learning which are often overlooked in the policy mobilities literature. The experience of BRT in South Africa reveals persistent introduction and alteration before adoption finally ensued; innovations spread across the globe through a series of unremarkable events and repeated suggestions that ultimately sanction it as a best practice to local policy implementers. Accordingly, I argue that failure is central to policy promotion by creating a process whereby learning is deliberate and delayed, assembling alongside earlier encounters to prolong the employment of international agencies. The book demonstrates that there is much to be learned by thinking through the practices of ostensibly ineffective, fruitless or aborted mobility. This contribution also raises questions about success and failure in policy mobilities, a topic of increasing importance among scholars.

Using Policy Mobilities as a Methodology

The physical, social and theoretical movement of ideas, objects, people, and places can be difficult to study because they are constantly in motion, positioning and repositioning, erratically and sometimes irrationally. Policies-in-motion demand mobile methods, a methodology "better suited to mobile times" (Clarke 2012). My research methods are therefore simultaneously "on the move" (Cresswell 2006), including physical travel on buses and between research subjects and sites, and "moored" (Hannam et al. 2006) in the offices of policy actors and in the operational BRT systems (Wood 2016). More than simply uncovering the movement, however, this study unravels the intricate connections and dependencies between ideas, objects, people and places, relationships essential for the acceptance of mobile knowledge (Büscher and Urry 2009; Urry 2007). This methodology resonates well with Law and Urry (2004) who reason that existing methods rooted in places insufficiently address fleeting temporality or transplanting places; that is that we must follow the chains, paths, threads and intersections in order to address the transitory nature of mobility.

For policy mobilities, two approaches – "follow the policy" (Peck and Theodore 2010a) and "follow the project" (Peck and Theodore 2012), which combine multi-sited ethnography (Marcus 1995) with the extended case method approach (Burawoy 1998, 2009) – are typically employed. Peck and Theodore (2012) suggest that the researcher (and their research) travel alongside the policy, concentrating on the power and politics of translation. Such a methodology, Peck (2011b) reasons, enables researchers to contemplate the role of policy networks and epistemic communities that form connections for enabling policy movement, experimentation and mutation across different cities. This methodology can be challenging, however, because as Peck and Theodore conclude, "it is not always possible to 'be here', when in the study of global policy networks there is a constant imperative to also 'be' somewhere else" (2012: 25). They therefore advise that researching mobile policies need not always be a multi-sited venture, but it will often necessitate "methodological travel, along the paths carved by the policies themselves" (2012: 24). Most importantly, the methods used should be sensitive to both the peripatetic nature of policy models and their affiliated policy actors as well as to the unpredictable character of espousal and emulation.

I study policy mobilities from the perspective of the adopting locality and examine the means by which a policy from elsewhere was introduced, shaped and localized by local policy actors, their interactions with global advocates and inter-referencing across space and time. Rather than observing or shadowing people's movements, as Marcus suggests in his understanding of "following the people", this study asks people to interpret and reflect on their own decisions and learning processes. Similarly, it does not simply "follow the thing" from Bogotá to South Africa per se; instead my perspective looks at how South Africans interpreted the mobility and assembly of BRT. This study "follows the mobility" by tracing the adoption of BRT from the assembly of policy model, through to the actors who first introduced BRT to South Africa, before focusing on the adoption process. This methodology draws on the experiences of the actors involved in the circulation and adoption of BRT, learning from their experiences and interactions with the South African version of BRT. It entails incorporating the site of origination with the site of adoption and the various stops along the way, all of which influence the uptake of a particular policy approach.

Previous studies following the products (Choy et al. 2009; Cook and Harrison 2007) and policies (Goldman 2005; Roy 2010), suggest a sense of completed transfer in which the learning has concluded and the adoption accomplished. But the story of BRT in South Africa is neither concluded nor accomplished. It therefore requires a retrospective trail – beginning with the adoption process and tracing back to the initial learning – to allow for reflection by policy actors, on the false starts, and slow decision-making that lead (or not) to BRT implementation. It is possible that the research process itself may have contributed further to the multifaceted and constantly mutating nature of policy mobilities: perhaps in the course of the interview, a policy actor may decide to change the trajectory of future emulation, or they may appreciate the influence of distantiated sites thereby enticing future learning. This supports McCann and Ward's (2012) call for researchers to move with the actors to understand how people, policy and place are made mobile. Rather than simply a physical movement between sites, this study moves temporally between the adoption of BRT and its initial arrival.

"Following the mobility" from the perspective of the adopting locality begins with interviews. Interviews are especially useful for probing beneath the sociopolitical exterior of the decision-making process. While interviews are sometimes criticized for being staged and scripted, especially when they involve educated and articulate elites (Peck and Theodore 2012), interviews are a relational process that expose not just the achievements of the adopting locality, but also the experimentation and failure associated with policy mobilities. They are also the best means for understanding the connections and disconnections between actors fueling the adoption process. This book developed inductively from their stories and experiences as well as their own learning process regarding BRT.

Interviews took place with senior planners and politicians from city (26 actors), provincial (2 actors) and national government (8 actors) as well as consultancies (26 actors), civil society (8 actors), academia (10 actors), donors (12 actors) and transportation operators (3 actors).[2] Of those interviewed, 68 percent were male, 32 percent were female; and 70 percent were White, 14 percent Indian, 12 percent Black, and 4 percent Coloured.[3] The meetings took place across South Africa from the Union Buildings and Parliament to the Civic Centre in Cape Town, the Metro Building in Johannesburg and the Transport Authority in eThekwini. More than two-thirds of meetings took place in either Cape Town or Johannesburg. Ten percent of respondents were located outside of South Africa – in Barcelona, Bogotá, Dar es Salaam, London, Manila, New York, Vancouver and Washington, D.C. – and these tended to be international consultants and academics. The majority of meetings were held in person at the participant's office. On average, the interviews took 1 hour and 30 minutes with the longest lasting 3 hours and 37 minutes and the shortest lasting just 20 minutes.

In selecting study participants, I was careful to avoid merely "studying up", by which Nader (1972) refers to studying the elites with power or "studying down", by asking the powerless. Instead, I "study through" (Shore and Wright 1997). This meant meeting with the actors who moved, shaped and adopted as well as those who opposed BRT. The list of interviewees included those currently involved in BRT introduction – for example each city's transportation officers, municipal politicians and engineers in the implementation agency – as well as those actors who have since moved on to other ventures – for example, from a city official to a private portfolio manager and from a minibus taxi driver to the CEO of a bus operating company.

I was also cautious against promoting agent-inflation, through which relatively unimportant actors become policy mobilizers by virtue of our discussion. I never asked a policy actor to tell me about their experiences with BRT. Rather, I asked respondents a few general questions about transportation to which they generally responded telling me about the BRT project, which included their experiences with policy mobilizers like Lloyd Wright and Enrique Penalosa, and on study tours to Bogotá. Philip van Ryneveld's role in bringing BRT to Cape Town, for instance, came up in several meetings across South Africa, none of which were prompted by me. I was careful to verify stories and experiences across interviews and with both internal and public documents.

Interviews were the primary methodology I adopted, but because that meant asking the individuals to describe their involvement, transcripts were triangulated with internal documents recording council meetings and study tours, as well as with public reports and presentations detailing implementation procedures. I reviewed more than a hundred important planning and policy documents found in the archives at both municipal and private planning offices, as well as frameworks (e.g. City of Johannesburg 2011), guides (e.g. ITDP 2017) and legislation (e.g. National Department of Transport 2009). I also evaluated advertising notices, architectural models, blueprints, brochures, films and websites. The urban successes "come alive" in clever policy documents and stunning showrooms (Pow 2014: 296). And seemingly mundane artifacts become "diverse technologies of seduction" that render policy models palatable for global consumption, thus revealing the socio-materiality of policy exchange (Bunnell and Das 2010).

Since BRT construction was ongoing and plans constantly emerging and changing, I also had opportunities to attend public meetings through which city officials publicized BRT, seminars wherein practical details were revealed to practitioners, and strategic meetings usually reserved for those directly involved with BRT. A public lecture by Gil Penalosa, Commissioner of Parks in Bogotá from 1999 to 2002, provided me with the opportunity to witness a policy mobilizer in-action as he introduced his perspective to a South African audience. Events such as this were often referenced in interviews – for example when Enrique Penalosa, former Mayor of Bogotá, or when Lloyd Wright, a global BRT advocate, presented the attributes of BRT to South Africans – as quasi-religious events full of animation and energy supported by high-resolution images of success. It was particularly informative to personally witness this style of knowledge exchange and have the opportunity to think about why it is so compelling.

All these methods require a willingness to move, but none are as mobile as the "go-along", which tends be either a walking interview (Carpiano 2009; Evans and Jones 2011) or a driving interview (Laurier 2004, 2008). The go-along accounts for the relationship between what people say and where they say it, and overall the process promises sounder results than sedentary interviews by prompting interviewees to connect with the surrounding landscape rather than the interviewer. Such methods reflect Simmel's (1950) perspective on social space, which he sees as the context for the creation of particular personalities and interactions (e.g. the stranger), and provides great insight into the learning process by allowing the researcher to examine participant's knowledge and experiences, as well as the way in which they engage with their social and spatial surroundings. The go-along is also particularly useful in reducing the typical power dynamics between the researcher and the subject.

My mobile methods included trips on the operational systems, tours of the bus depot and control center and driving tours across the city. Ordinary journeys on Cape Town's MyCiTi and Johannesburg's Rea Vaya usually included conversations with drivers and passengers. One driver told me how much he misses driving a taxi because he has to work many more hours as a Rea Vaya bus operator, while another prefers the regular employment and guaranteed salary. One of the more extraordinary mobile interviews was a driving tour with Ron Haiden, then-Manager of Infrastructure and Development in Cape Town. We drove for hours visiting and discussing transportation systems and services across Cape Town. I learned a great deal from seeing the projects first-hand and I enjoyed hearing anecdotes about "The Wright and (w)Ron(g) Show" (when Haiden worked with Lloyd Wright). Haiden's insights into both the details of station design as well as the big picture needed to conceptualize Cape Town's entire My-CiTi system, were extremely valuable in situating BRT within the broader efforts at transportation transformation.

My professional experiences working with South African cities also provided me with the foundations from which to launch this research. Between 2008 and 2010, I worked as Programs Manager for the South African Cities Network (SACN). In this capacity, I was responsible for a variety of projects, programs and publications, including managing exchanges between cities implementing BRT systems. One interesting project was a learning event hosted jointly with Johannesburg in October 2009, in which the City invited other South African cities to ride the new BRT system and learn from Johannesburg. My experience with the distribution of BRT makes me a policy actor and thereby has enabled me to draw on my own experiences, while simultaneously having access to empirical knowledge of the policymaking process. These experiences empower me to "reconstruct a landscape in the eyes of its occupants" and to imagine the experiences of policy actors (Samuels 1981: 129). Although I did not occupy a formal position after 2010, I acknowledge my role as a formerly active policy actor entering into circulation processes with local actors.

Structure of the Book

How Cities Learn is organized into seven chapters including this introduction. Chapter 2, "Geographies of knowledge", provides the analytical scaffolding for this book. It introduces the policy mobilities framework which will be used to explain the multiple and complex relations between actors, which elevate the achievements of particular policies and bring certain cities into conversation with one another, while pushing other ideas, policymakers and places further apart.

Chapter 3, "Translating BRT to South Africa", traces the global geography of BRT to understand what features attracted South African policymakers and, in so doing, reflects on the theoretical notion of best practice policy models and their process of mobilization, mutation and translation. Moving away from the assumption that BRT is merely a technological template for stations and busways, this chapter reveals that the Bogotá model was not replicated in South African cities but rather that a variation, which focused on the transformation of the informal transit system, emerged. This suggests that policy models are polysemic, and rather than being duplicated they are modified and mutated to suit the needs of each importing jurisdiction. The adoption of best practice therefore takes place only when aspects of the circulating notion combine with underlying conditions, thereby normalizing the particular technical arrangements and social rationalities for adoption.

Chapter 4, "Actors and associations circulating BRT", focuses on the urban planning professionals, practitioners and politicians introducing, circulating and managing the adoption of BRT. While some are instrumental in planting ideas that may lie dormant for some time, others engage in their prospective evaluation and actively stimulate their application. International policy actors cannot simultaneously create, impart, mobilize, and approve global policy models. Instead, it is the local (South African) policy actors that localize international best practice. Their interactions with internationals legitimize global policy by giving it both local and transnational salience. This examination of the policy actors expands our understanding of the varied direction, speed and influence of global and local influences shaping the contemporary city.

Chapter 5, "The local politics of BRT", analyzes the international, national and local connections and disconnections between localities that influenced its adoption. At the international scale, this chapter reveals a deliberate preference to learn from Bogotá rather than a multiplicity of South American cities who also implemented BRT. This same enthusiasm for south–south exchanges was also used to disregard the experiences of African and Indian cities. Within South Africa, this chapter explores the competitive political and technical relationships between cities that influenced the adoption of BRT. This multi-scalar analysis of the politics of BRT explains the process by which policy and policy actors connect and disconnect topographically and topologically.

Chapter 6, "Repetitive processes of BRT", situates BRT adoption within a longer history of South African transportation planning. It exposes previous involvements with BRT-like interventions that did not progress. This chapter contests the fast policy literature, which identifies the introduction of prefabricated best practice policies as part of the shortening of the policymaking cycles. Rather it suggests that there are multiple temporalities through which circulated policies emerge and remerge before adoption, and that often without these multiple attempts, policy circulation would not be effective. BRT learning is therefore gradual, repetitive and at times delayed.

Chapter 7, "Conclusion", outlines the book's main theoretical arguments, in particular answering how and why cities adopt circulated forms of knowledge. Here, I argue that policy mobilities is a process of learning and understanding, adoption and adaption, competition and collaboration, facilitated by local South African policy actors and their relations with urban elsewheres. In concluding, *How Cities Learn* explores the impact of BRT on the socio-spatial landscape of the South African city, and situates this study within the wider process of post-apartheid transformation.

Notes

1 Interviews 69 and 62.
2 Appendix A includes a list of interview respondents including their title, organizational affiliation, place and date of the interview. Appendix A is used throughout the book to link interview material with interviewees through a numerical system that lists the interviews chronologically. A number in brackets (e.g., [15]) refers to a particular interview and the reader should turn to Appendix A for supplemental information on that source.
3 Racial categories are a legacy of apartheid in which all South Africans were defined according to these four classifications. These terminologies continue today. I have included the racial categories of my interview respondents to inform the reader of the extreme disproportionality between those planning for transport and those using the transport.

Chapter Two
Geographies of Knowledge

Building an Analytic for Tracing

The widespread adoption of BRT in South Africa denotes a process of "policy mobilities" in which localities create, circulate and adopt global innovation (McCann 2011b; McCann and Ward 2011). *How Cities Learn* reveals that policy mobilities is not only the "purposeful, repetitive, programmable sequence of exchange and interaction between physically disjointed positions held by social actors in the economic, political, and symbolic structures of society" (Castells 1996: 442), but also a process of dialogue and debate which involves power and personalities. These practices, however, have proven difficult to research since the exchanges rarely lead directly to uptake. Previous studies have traced the movement of knowledge through various "coordination tools" (McFarlane 2011b: 364), which include consultancies (Rapoport and Hult 2017; Wood 2019b), conferences (Cook and Ward 2012; Temenos 2016), study tours (Montero 2016; Ward 2011; Wood 2014a), technology (Rapoport 2015), and workshops (Wood 2014b); and others have followed the transnational advocacy groups (Stone 2002) and learning organizations (Wood 2019c) that package, frame and legitimize global circulation (Theodore and Peck 2011). Scholars tend to conclude that learning emerges through various voices, interests and expectations, translating and coordinating a multitude of information, including existing knowledge, across asymmetrical power structures and creating possibilities from the impossible (McFarlane 2011a). *How Cities Learn* builds from this scholarship by outlining a conceptual and practical analysis of policy mobilities that attends to the plethora of ordinary practices – be it through engagements with fellow practitioners, with their toolbox of material solutions, or after a particular moment of discovery – that form the assemblages of learning.

It is along this line of inquiry that *How Cities Learn* builds upon and extends the thinking of urban scholars, first by bringing concerns of power into question, and second by problematizing notions of governance at-a-distance. Certainly policy mobilities arguments have provided evidence that power is now disseminated across a host of diverse agents and agencies. In this book, I ground policy mobilities within the adopting locality so as to suggest that power is always exercised in situ, although in the case of policy mobilities it often seems as if power is furthered by external authorities. Indeed, policy mobilities is a practice of both embracing extra-territorial thinking, but equally so a means through which local actors exploit international advocates and their policy models to justify preordained decisions, which might otherwise be resisted by local politics. Thus policy mobilities, though global in nature, is an inherently local process, one that is best exposed by scrutinizing the actors and their actions within the adopting locality, and then tracing back through their rationale for implementing a policy, product or practice also found elsewhere.

Accordingly, I employ the process of "tracing" to better understand policy mobilities (Wood 2020). This approach draws on Robinson's three genetic and generative approaches to comparative urbanism: "composing" – that is examining the specific similarities and dissimilarities within a range of instances; "launching" – that is starting from anywhere and then putting the analysis to work anywhere; and "tracing" – that is following (i.e. the genetic component) and comparing (i.e. the generative component) the connections themselves. It involves outlining the connections and their influence on the comparable instances (McCann 2011b; Ward 2006), as well as comparing cities and their relationships themselves (Myers 2014; Söderström 2014). This approach allows us to trace historical events and consider their interrelated effects on the urban (Cook et al. 2014a; Wood 2015a); it enables us to see the urban realm as an assemblage of the here-and-there (McCann and Ward 2011); and it supports further consideration of the interrelatedness between cities within this stretched and extended moment of urbanization (Roy and Ong 2011). This means not only tracing that which brings cities into conversation with one another (i.e. the presence of comparativism) but also that which does not bring cities into conversation (i.e. the absence of comparativism), as well as the inherent subjectivity and slipperiness of those relations.

This chapter introduces the multidisciplinary approach of the policy mobilities literature and explores the research thematically, in the process also setting up a structure for the remainder of the book. First, I question the role of the innovation itself in attracting attention from the adopting communities – that is how and why are certain mobile ideas attractive to importing localities, and is there an agency inherent to certain policy models? Second, I refocus on the policy actors assembling, mobilizing and adopting innovation, to draw attention to their role both individually and within networks in the adoption of circulated forms of knowledge. Third, I concentrate on cities as institutions and specifically the way in which wider municipal-level decisions and relationships influence adoption procedures. Fourth, the argument turns towards temporality, to understand the way in which previous experience with similar forms of knowledge enables a more rapid adoption when the importing locality is ready. These discussions inform the key findings of this book.

Tracing through Policy Models

Policy mobilities scholarship studies how policies move. A geographically based understanding extends beyond merely classifying the thing moving and instead considers how policy models travel and what happens along the way (Peck and Theodore 2010a; McCann and Ward 2011). A brief review of the etymology of "policy model" presents it as a "complex social construction (no less 'real' for all that) which can only be understood by studying both its apparent 'internal' characteristics and simultaneously, its 'external' relations, which are co-constituted" (McCann and Ward 2013: 4). The term "model" is generally defined as "a way of doing something from which others learn" (McCann and Ward 2013: 3). McCann (2011a) uses the terminology "policies" to describe the formally drafted guidelines for governance, "policy models" for the statements of ideal policies and "policy knowledge" as the expertise about good policymaking and implementation. Policy mobilities theorists use a neo-Foucauldian lens to frame the role of governmental technologies and rationalities in shaping contemporary urban–global political economies. They emphasize the seemingly mundane techniques, discourses, representations, and practices through which policy models are made up. For the purposes of this book, a policy model is any form of circulated knowledge perceived to be affiliated with another city, individual or government that when introduced is assumed to be an improvement over its predecessor. Such a broad definition is illustrative of my overall approach, which emphasizes the relational co-production of the material and social aspects of local and global policymaking.

Previous studies have considered the Barcelona model of urban regeneration (González 2011), the Austin model of creativity (Florida 2002; McCann 2004), the Porto Alegre model of participatory budgeting (Crot 2010; Porto de Oliveira 2017), the New York City model of business improvement districts (Hoyt 2006; Ward, 2006, 2007a, 2007b, 2011), and the Vancouver model of sustainability (McCann 2008). Evidence suggests that these models are replicated because of a strong symbolic association with their place of origin that supersedes their particular achievements. The Singapore model, for instance, is not a set of concrete procedures for reproducing social housing or sustainable urbanism but a brand synonymous with efficiency and success (Hoffman 2011; Chua 2011;

Ong 2011). Chua (2011) makes a case for how the urban planning innovations in Singapore are packaged and marketed as an exportable brand and while it has become a global blueprint for sustainability, it is merely a symbolic association because the achievements in Singapore are firmly rooted in the socio-political location and therefore cannot really be reproduced elsewhere. And Hoffman (2011) suggests that the use of the term sustainability has become a new criterion for ranking cities in general – those that are sustainable are more attractive, completive and viable than those without this nomenclature – a point I later make to explain the magnetism of BRT. Perhaps then models are not merely a set of guidelines to be duplicated but part of a wider approach to planning which can be inserted into complex political conditions. Those cities that implement a best practice policy model are part of a cohort of livable and sustainable cities, and those who do not are not.

Sometimes policies are only adopted after they have been reshaped to suit the needs of the importing locality. A pivotal study by González (2011) looks at how urban regeneration models travel and mutate, by considering the influence of Bilbao and Barcelona to fit the needs of a diversity of adopting localities. These cities abstract their locally developed plans and programs by building complex systems to package the model through workbooks and the requisite set of site visits needed to promote themselves as role models. Some elements of the exported model may not be suited for the importing locality, so it is adapted for its new context. Robinson (2011) concludes then that policy models are always subject to negotiation as they are circulated and practiced at the local level. Yet, relatively little is known about how these processes of mutation take place. The discussion in Chapter 3 adds a critical contribution to the policy mobilities field, by revealing the way in which models mutate as they travel, at times becoming stylized versions of themselves.

Tracing the origins of a policy model reveals that adopting localities are attracted to the city-wide outcomes promised to accompany the model. Business improvement districts for example only become policy models and move elsewhere because their achievements are visible. In this case, the maintenance, promotion and security services result in cleaner, safer commercial districts, achievements easily advertised to needy property owners, developers and government actors in localities in another city. In Chapter 3, I argue that policy models move because of their association with elsewhere, which enables their reproduction in spite of different geopolitical conditions in the importing locality. My approach casts doubt on the extent to which a model can be faithfully duplicated elsewhere, and instead considers the ways in which its immaterial features encourage locally contingent policymaking, loosely based on the achievements elsewhere.

Such an approach leaves open an opportunity to think about policy models as mobile objects, packaged and produced into forms amenable to travel, and circulated by global and local actors who carry them from city to city, translating and applying them. They are imported by ambitious places eager to replicate the achievements elsewhere. In Chapter 3, I explore the agency of policy models and their role in forwarding their own circulation and adoption. How did BRT spread around the world, and why was this particular model adopted in so many cities? In answering these questions, I explore how policy packages spread not because they work but because they conform to extant power structures and established interests. Thus, a successfully transferred policy is not evidence of the superiority of any particular practice or policy, but rather indicative of the particularities of locality and its policy actors (Chapter 4) as well as relational connections between cities (Chapter 5).

Chapter 3 traces the global geography of BRT to understand what features attracted South African policymakers to this international best practice, and in so doing reflects on the theoretical notion of policy models and their process of mobilization, mutation and translation. Moving away from the assumption that policy models provide a technological template for best practice reproduction, this book considers the way in which the policy introduced differs from the one at its site of origin. The empirical focus reveals that Bogotá model of BRT was not duplicated in South African cities but rather a variation, which focuses on the transformation of the minibus taxi system emerged. This suggests that policy models are polysemic and rather than duplicate, they modify and mutate to suit the needs of each importing jurisdiction. The adoption of best practice therefore takes place only when aspects of the circulating notion transform and combine with underlying conditions, thereby normalizing the particular technical arrangements and social rationalities for adoption. In concluding Chapter 3, it will be evident that human actors are central in the circulation and adoption of policy models, considerations that are the subject of the next section.

Tracing through Actors and Associations

Building on the debates around traveling urban policies, I argue that people and their associations drive learning and policy mobilities. Policy actors, including both state and non-state actors as well as architects and engineers (Rapoport and Hult 2017), consultants (Wood 2019b), philanthropists (Stone 2010) and so forth, frequently look elsewhere to detect, decode and distribute best policy practices (Cook 2008; Dolowitz and Marsh 2000; McCann 2008, 2011a; Theodore and Peck 2001; Wolman 1992; Wolman and Page 2002). The background, experiences, economic, and political position of each policy actor influence their decisions. Therefore, explorations of agency and power are important in understanding the way in which global and local actors survey and circulate particular policies.

Two scholars of the operation of global consultancies provide useful groundwork for this study of BRT. Larner and Laurie (2010) investigation of traveling technocrats in New Zealand and McCann's (2011b) study of Bing Thom, a

Vancouver architect-consultant hired by Fort Worth, Texas to "Vancouverize" the city, offer insight into the central role of non-state, extra-local actors in facilitating policy mobility. Larner and Laurie illustrate the role of traveling technocrats in the movements to globalize and privatize telecommunication and water services. In their analysis of mid-level technocrats enacting privatization techniques, they note that there are a number of supplementary actors including, in their case, New Zealand telecommunications and British water engineers mobilizing transnational flows. These professionals supply knowledge and cultivate its local adoption, forming associations that advance policy mobilities. McCann underscores the ability of policy mobilizers to influence development across localities, in spite of, and because of their extraterritorial authority. McCann (2013) furthers these arguments in his study of "policy boosterism", a concept he uses to understand how policy knowledge is mobilized around the globe. He suggests that Vancouver's extrospection is tied to a range of global and local policymakers. These contributions highlight the way in which localities are influenced by intermediaries in the circulation of best practice.

Also of interest is a study by Ward (2011) that offers insight into the actors and their methods for employing policy models in Manchester, UK. Their research focuses on Manchester city officials pursuing knowledge of Olympic and Commonwealth Games projects, and specifically the techniques in which they learned from previous host cities to exploit these opportunities to engage in broader economic development strategies. Officials visited cities whose best practice in hosting mega events had been broadcast around the world, such as Los Angeles and Lillehammer; their meetings with officials as well as site visits to sports-related infrastructure and associated regeneration sites were used to execute the redevelopment of east Manchester. In this case, the city's failed bid to host a mega-event did not deter them from implementing the learning. Such points offer an opportunity to consider the manner in which local politics is used to convince (or subvert) particular actions and determinations taking place elsewhere as a part of the learning process. It also provides evidence of the importance of previous policy failure in ongoing decisions.

Such considerations for global and local policy actors, raises questions around who holds power in the city and who determines urban policy? Furthermore, how do policy mobilizers exert normative power albeit spatially and politically distantiated? Moreover, if international players are acting as decision-makers, then how do local elected officials such as mayors wield power differently than civil servants? Finally, how does the shifting role between public and private enterprise enable those outside of government to exercise power? A Foucauldian understanding of power suggests that it is something which circulates and "not only do individuals circulate between its threads; they are always in a position of simultaneously undergoing and exercising power" (1980: 98). Power thus is an "entangled bundle of exchanges dispersed 'everywhere' through society" (Sharp et al. 2000: 20). It is not composed of a fixed group acting in unison, but rather it is diffused through a multitude of places and spaces through the circulation of knowledge. Power is not limited or enhanced by geographic distance because actors can reach into other localities, influence decisions without political power and even keep ideas out of foreign localities (Allen 2003). The arguments presented in Chapter 4 do not presume that power is situated within institutions as Castells' (1996) "space of flows", or Lefebvre's (1991) description of the systematic reorganization of institutional geographies, but rather follows Foucault's suggestion that power is not a "zero-sum game", an ensemble of actions that overlap and support one another to achieve mutually beneficial gains (Dreyfus and Rabinow 1982).

Chapter 4 explores the varied and sometimes unexpected decisions by policy actors as expressions of power relations, as well as the relationships formed prior to and because of the circulation process, which facilitates the adoption of ideas from elsewhere. This includes some reflection on actor-network theory and an exploration of both human and nonhuman actors within policy mobilities. In Chapter 4, I draw on Deleuzian ideas of assemblages to theorize and explain how networks form and sustain these circulations as part of network formation and assembly (Deleuze and Guattari 1972, 2004; Foucault 2003; Rabinow 1984). This theorization places policy travels and transfers alongside a host of urban happenings – capitalism, political contestation, as well as learning and mobility (Farias and Bender 2010; McFarlane 2011b). Larner and Laurie (2010) show how, within these global assemblages, experts easily travel between places mobilizing transnational flows (see also Larner 2002). Both McFarlane's (2009) study of transnational movements and Robins' (2008) research into social movements use this approach to demonstrate how networks exchange experiences, ideas and resources across the globe. McFarlane (2011a) interprets the movement of knowledge across spatiotemporal circumstances through the notion of assemblage, in which the actors and their knowledge and materials form an agglomeration. It is important to realize though that the assemblage is temporary – it constantly adjusts and alters over time (Ong and Collier 2005).

My exploration of both the global and local individuals involved in policy mobilities in Chapter 4 enriches these arguments, with empirical examples of international advocates and local implementers adding a focus on the intermediaries who link these two sets of actors, through exploration of their interactions. This focus on a wide variety of actors is particularly critical in understanding how and why circulated ideas are adopted locally. The information and experiences presented by these policy mobilizers are validated by virtue of their presence in the locality. That is, they do not simply land elsewhere; intermediaries, elected officials and civil servants in need of their expertise usually invite them or choose to provide an audience. The policy mobilizers usually present a vast field of possibilities but subtly direct their hosts towards a preordained solution. The local actors also participate in the game – they listen to the presentations, ask questions and at times are critical of the talk, before ultimately arriving at the same conclusions as the policy mobilizers. In some instances, the local actors rely on these international voices to validate previously planned urban investments. Chapter 4 will demonstrate the way in which policy mobilizers and local actors maintain a symbiotic relationship, both exploiting the other's positionality to adopt circulated policy.

Tracing through Cities

The relationships between cities play a central role in this study. A range of economic and political characteristics bring certain cities into conversation with one another, while pushing others further apart. The learning process often requires localities to work closely together, sharing private technical and political information and, in many instances, spending extended periods collaborating. Peck and Theodore (2001, 2010b), for example, present instances of the transfer of poverty alleviation policies between the UK and the US during the Thatcher/Reagan and Blair/Clinton eras, arguing that the friendship between the political leaders predisposed the governments to collaborate. The manner in which commonalities in government and policymaking contexts enable policies to transfer across socio-political boundaries is also evident in studies of the transnationalization of the business improvement district, where geographical proximity between the UK and Europe was disregarded in favour of exchange of political associations (Ward 2007a, 2011). There have also been studies of similar accounts of ideological exchanges between the former Soviet countries (Cook et al. 2014b; Offe 1996), as well as between European cities after the Second World War (Clarke 2010; Vion 2002). In such instances of "municipal diplomacy" (Saunier 2002: 526), cities are not merely importers or exporters of policy but part of the global system of power relations in which policy circulates.

Cities at times also search for new ideas beyond their most obvious comparators, perhaps assuming that there is more to appreciate from engaging with difference. Mahon and Macdonald (2010) present a case study comparing poverty alleviation programs in Toronto and Mexico City. In this instance, the two cities exchanged ideas particularly because their different approaches offered innovation to each city. However, in exchanging local solutions, Toronto and Mexico City strengthened their relationship, thereby demonstrating that policy exchange is instrumental in forming relationships between localities. In McCann's (2011b) study of Bing Thom, a Vancouver architect-consultant hired by Fort Worth, Texas to "Vancouverize" the city, he underscores the ability of a policy mobilizer to influence development across localities despite considerable differences between the places. These policy mobilities studies provide a foundation for my research of the manner in which cultural, economic and political relations between and within cities were used by South African policymakers to advance (or subvert) particular actions and decisions regarding BRT.

It is also important to note the imbalance of power between cities of the global north and south (Massey 2011; Robinson 2011), as well as across cities of the global south (Bunnell and Das 2010 for urban policy transfer from Kuala Lumpur to Hyderabad, and Hains 2011 for transfer between Dubai and Delhi). Massey (2011), for instance, outlines a trade agreement between London and Caracas whereby London provided technical advice on a range of urban issues from transportation planning to waste disposal in exchange for a reduced price on oil, which London then used to fuel the city's buses, funding a 50 percent reduction in fares for the poorest people of the city. The relationship between the cities became the subject of contestation in 2008 when the newly elected Mayor Boris Johnson cancelled the agreement. This account provides evidence of the "politics of place" as the "outcome of the contested negotiation of physical proximity", which is also "explicitly relational" beyond the confines of the city (2011: 4). Such a contested terrain is also found by Robinson in her analysis of city development strategies (2011). She looks at the circulation of city strategies as a technique for governance, introduced in Johannesburg and then disseminated by the World Bank and Cities Alliance. "What might appear to be an instance of the local application of global policy discourse in the Johannesburg case", Robinson writes, "was a strongly locally determined policy process, shaped by quite specific political dynamics" (2011: 34). Such claims offer a counterhegemonic view that policy practice can always be traced to foreign or western origins, a point critical to my study, which recognizes the multiplicity of sites of origin that are included or disregarded in the circulation of best practice.

How Cities Learn considers how policy exchange and adoption connects places. Mobilities theorists have provided a variety of metaphors for interpreting the way in which a policy moves through a system – the train moves along the tracks (Latour 1993, 2005), a car at a petrol station (Normack 2006) or water in a creek (Tsing 2000), all of which presuppose the relationship between the object and the conditions of its mobility. Tsing (2000: 337) reasons that "a focus on circulations shows us the movement of people, things, ideas, or institutions, but what it does not show is how this movement depends on defining tracks and grounds or scales and units of agency... If we imagined creeks, perhaps the model would be different; we might notice the channel as well as the water moving". These entities – the train, car and water – circulate through the system while maintaining their connection to the spatial elements – the track, petrol station and creek bed. These movements underscore the "networked nature of interconnectedness" (Lester 2006: 135). Even though the relationship between the movable objects and place is provisional, ephemeral and transient – the train moves to a different part of the track, the car to another petrol station and the water through the creek – it remains linked through the system. This focus on intergovernmental mobility considers the way in which the (metaphorical) tracks, petrol station and creek bed promote circulations by creating a conducive context for their continuity. Neither the terrain nor its connections are "spatially fixed geographical containers for social process" (Hannam et al. 2006: 5), but these metaphors can help us understand the inter-urban and intra-urban relationships that facilitate the exchange of BRT across localities.

Chapter 5 examines the relationships between importing and exporting localities. It investigates the influence of municipal politics that enable and/or constrain decisions to adopt – that is the way in which those cities learning concurrently are connected and disconnected by these multifaceted processes of knowledge accumulation. Such debates deepen and widen the space through which policy flows, by proposing that local municipal relationships, both competitive and cooperative, shape the circulation process. Moreover, it exposes policy mobilities as more than a course through which energetic policy mobilizers introduce proven solutions to unsuspecting policymakers. Rather, policy ideas move through the terrain of local politics, which house prevailing international and domestic cooperative and competitive relations.

Tracing through Temporalities

How Cities Learn considers the role of temporality and historicity in policy mobilities. It illustrates that learning is a progressive conversation in which slowly, incrementally policymakers warm to an otherwise foreign notion and through these continuous exchanges, ideas, practices and programs are relocated and localized. This research will take a broader perspective to show that policy flows through "waves of innovation" (McCann and Ward 2010: 175) which only seem to be arriving more frequently. There are peaks – periods of rapid diffusion facilitated by either need or opportunity, or both – and valleys – periods when circulation is minimal. In South Africa, the learning about BRT was longwinded and drawn out, incremental and at times delayed. *How Cities Learn* illustrates that regardless of the speed of circulation, policy implementation remains cumbersome because policy is always political, meaning that it takes time to localize policy.

Arguments supporting a crisis-driven approach towards policymaking through the hurried acquisition of international policy models, are frequent in the policy mobilities literature (Brenner et al. 2010; Brenner and Theodore 2002; Clarke 2009; Peck 2002, 2003, 2011a; Peck and Theodore 2001; Theodore and Peck 1999, 2001, 2011; Ward 2011). The term "fast policy" (Jessop and Peck 1998; Peck and Theodore 2015) has been used to characterize the rapid introduction of off-the-shelf prefabricated best practice policies. Much of the fast policy discourse rests on the prevalence of knowledge sharing between the UK and the US in the 1980s and 1990s. This scholarship proposes that explosive tactics create momentum to drive through systematic transformation, suggesting that a more incremental approach could not sustain implementation procedures (Peck and Theodore 2010a, 2010b). In so doing, it takes for granted these processes as hectic and hurried arguing that ideas and innovations are appropriated because of their prevailing success elsewhere, which make them easily executed within local policymaking cycles.

Departing from the prevailing logic in the fast policy literature, in Chapter 6 I demonstrate that the learning process is in reality often lengthy, incremental and at times delayed. The dissemination of Singapore's electronic road pricing system for taxing vehicular movement through the city provides an illustration of the multiple temporalities of policy learning. Whereas in London, a version of the pricing system was appropriated successfully, in New York, the Mayor's proposal was defeated by the state legislature (Chua 2011). These arguments illustrate a difference between the speed at which a policy moves and the velocity at which it is implemented. Unlike Kingdon's "policy window" (1995) which assumes a random confluence of people, choices, problems and solutions that come together at a particular juncture to enable learning (and if that window closes, the opportunity is missed), my findings suggest that ideas are acted upon through multiple temporalities. Policy adoption is an inherently political process and thus the speed at which a policy is adopted is distinct from its measures of long-term efficacy. The achievements of a new transportation system, be it related to its financial viability or impact on the city, may take two decades to become apparent and thus divorced from their assumed likelihood of success at the time of adoption. It seems then that there are a number of different speeds and temporalities through which policy flows.

Chapter 6 considers the multiple temporalities through which circulated policies emerge and remerge before adoption, reasoning that without these multiple attempts, policy adoption is unlikely to occur. The arguments that follow bridge the lacuna between historical and policy mobilities studies, illustrating the gradual, protracted and idiomatic manner through which transnational best practice proceeds, in contrast to the spontaneous and hasty process documented in scholarly literature on policy mobilities. Regardless of the speed through which best practice spreads, policy application remains cumbersome because policy is inherently political, involving people and personalities as well as regulations and restrictions. While it may seem as if circulated policies shorten the gestation time from policy introduction to policy adoption, these repeated attempts ensure that the turnover only appears accelerated.

The remainder of the book considers these issues in further detail.

Chapter Three
Translating BRT to South Africa

Introduction

BRT systems currently operate in nearly 160 cities. With 99 of the cities initiating new systems since 2001 for a total of 280 corridors and 4,300 km of busways and moving 28 million passengers per day, it is one of the fastest growing transportation technologies in the 21st century (ALC-BRT and Embarq 2013). While this impressive growth can be attributed to the achievements of its physical features and the services it provides, this chapter focuses on the BRT as a policy model to understand why so many policymakers from such a diversity of localities have implemented it. This chapter examines those features of BRT that attracted South African policymakers and, in so doing, reflects theoretically on policy models and their process of mobilization, mutation and translation. For policymakers in South Africa, BRT is not merely a "technical option" [62] or concrete, prescriptive municipal or national policy but part of "a series of projects … linked to other transport and land-use projects" [62]. My argument is that the mutable nature of policy models – that is their ability to transform into whatever the importing locality needs, which in the case of BRT manifested in the South African city as both a transportation technology and a strategy for post-apartheid transformation – facilitates its universal application in a diversity of pioneering contexts. Local planners and politicians were drawn to BRT precisely for its mutability, recognizing BRT as not only a tool for mobility but also a means for economic and social urban transformation. The South African experience therefore is more than just an imitation of the tubular stations in Curitiba or red busways in Bogotá, it is the story of comprehensive local change.

How Cities Learn: Tracing Bus Rapid Transit in South Africa, First Edition. Astrid Wood.
© 2022 Royal Geographical Society (with the Institute of British Geographers). Published 2022 by John Wiley & Sons Ltd.

The aim of this chapter is to examine BRT as a policy model and trace its circulation and mutation from the site of assembly to place of adoption. Policy models like BRT, it seems, are not only on the move but can take on new forms through the transfer process. In telling the story of BRT adoption in South Africa, this chapter examines what is actually moving through the circulation process and how a project's materiality and immateriality shape its circulation and adoption. This "preferred bundle of practices" (Peck and Theodore 2010a: 171) is crystallized into a particular set of procedures and conventions, and is not necessarily duplicated by the importing to a locality. I argue that models are ascendant for both their material and immaterial features: in the case of BRT, the buses, bus lanes and platforms are the material representation of the model, while the ability to incorporate affected informal transit operators and validate municipal centrality in the transportation function represent the less tangible features, which in many instances are more attractive to the importing locality. Both material flexibility and representational valence are necessary for the movement of a policy model and this aids the simultaneous mutation of and adherence to the model. This flexibility makes BRT acceptance possible in spite of considerable variations between espousing localities. These contentions made in this chapter suggest a previously underexplored aspect of the agency of policy models, assuming that certain models are superior to others not because they are exceptional but because they are readily translatable.

The next section defines the attributes of BRT most often replicated and references the many cities that have introduced them. From there, this chapter concentrates on the realization of Transmilenio in Bogotá and its subsequent mobilization. This leads to a more detailed presentation of BRT adoption: first, outlining the procedures through which each of the six cities of the study learned of and adopted BRT; and second, detailing the physical characteristics of the various BRT projects. The final empirical section evaluates the process of taxi transformation, including a brief analysis of the history of the industry and its relationship with government. In concluding, it will become evident that the successful introduction of mobile knowledge takes place only when aspects of the circulating notion combine well with underlying local conditions, thereby rationalizing the particular technical arrangement.

The Geography of BRT

Much has been written about the accomplishments of BRT in Latin America (Beltran et al. 2013; Delmelle and Casas 2012; Hidalgo and Graftieaux 2008; Hidalgo and Gutiérrez 2013; Hodgson et al. 2013; Marsden et al. 2011; Marsden and Stead 2011; Mejía-Dugand et al. 2013), as well as in Australia (Bray et al. 2011), China (Deng and Nelson 2013; Matsumoto 2007; Zhang et al. 2014; Zheng and Jiaqing 2007), India (Kumar et al. 2012; Ponnaluri 2011), Indonesia

(Matsumoto 2007), Korea (Cervero and Kang 2011) and the United States (Cain et al. 2007; Callaghan and Vincent 2007). In South Africa, researchers have considered the influence of BRT on transportation planning (Kane and Del Mistro 2003), poverty reduction (Venter and Vaz 2011) and personal travel behavior (Behrens and Del Mistro 2010), amidst concerns for formalizing the affected transit operators (Schalekamp and Behrens 2013; Venter 2013), institutional realignment (Walters 2013) and the viability of the system (Salazar Ferro et al. 2013). This chapter presents an entirely new argument, departing from previous accounts in the transportation literature, which typically attributes the replication of BRT to its technical merits that enable policymakers to easily and quickly employ a cheap, comprehensive solution (Deng and Nelson 2011; Gilbert 2008; Hidalgo and Graftieaux 2008). Instead, I focus on the social and political processes driving the replication of BRT around the world.

There is no precise definition of BRT. It is a concept of high-quality bus-based transportation that provides affordable, comfortable and speedy service through the provision of segregated infrastructure, rapid and frequent operations and customer service (Wright 2007a). BRT is described as a "flexible, rubber-tired form of rapid transit that combines stations, vehicles, services, running ways and information technologies into an integrated system with strong identity" (Levinson et al. 2003). Such broad characterizations produce a spectrum of interpretation ranging from "BRT-lite", in which some form of bus priority is developed but the system is not fully segregated to "full BRT", which provides metro-quality service, integrated network of routes and corridors, closed stations, and preboard fare collection. These systems improve upon conventional bus services, which tend to move through mixed traffic and utilize on-board fare collection as well as basic busways, which may operate in segregated busways but still use on-board fare collection. The circulated model of BRT generally includes segregated busways for use only by official operators, an integrated network of trunk and feeder services, stations that enable preboarding fare collection and fare integration between routes, and a large fleet of articulated buses managed by an intelligent transportation system (ITS), supported by business, marketing and operational plans.

The concept of BRT can be traced back to 1937 when Chicago outlined plans for express bus corridors, and the first mention of the term BRT was in a 1966 study for the American Automobile Association by Wilbur Smith and Associates (Levinson et al. 2003). Although Lima was the first South American city to realize the notion with its Via Expresa, which began operating in 1972, Curitiba's Rede Integrada de Transporte, which opened in 1974, is reported to be the first complete BRT system (Mejía-Dugand et al. 2013). Many of these cities learned from their predecessors: Bogotá-based technical and political teams visited Curitiba to learn about BRT (Hidalgo and Graftieaux 2008); following a 1998 visit to Curitiba, officials in Los Angeles proposed building a BRT line along an abandoned railway line in the San Fernando Valley (Callaghan and Vincent 2007);

in February 2003, a delegation of 15 Indonesian representatives attended the International Seminar on Human Mobility in Bogotá and, when they returned, established a task force to launch a BRT system in Jakarta (ITDP 2014, 2019). This learning was recurrent and continuous, as evidenced in the retrofit of off-board fare collection and tubular stations along Curitiba's exclusive bus lanes in the 1990s (Hook 2014). In some cases, international organizations play an important role in terms of funding. In Bogotá, the World Bank was one of the funding sources for infrastructure and the Spanish Development Fund funded the total cost of the first 11.2 km of Quito's trolley line (Matsumoto et al. 2007).

Most BRT systems have greatly improved mobility by reducing travel time and improving comfort and reliability. The indicators of system performance, however, differ from city to city: whereas Bogotá's Transmilenio has been celebrated for its peak loading (passengers/hour/direction), Istanbul's Metrobus has been celebrated for its commercial speed (kilometer/hour), the Guangzhou BRT for its infrastructure (passenger boarding/kilometer of busway), and Guayaquil's Metrovia for operational productivity (passenger boardings/bus-kilometer) (Hidalgo and Gutiérrez 2013). Ridership figures also vary enormously: from 270,000 passengers per weekday on Sao Paulo's Interligado to 80,000 in Quito (Hidalgo and Graftieaux 2008). Because there is no single definition of BRT, variations in the physical characteristics of the system include the location and design of busways (median or curb side), the type of platform (high or low floor), the routing (radial or diametric, short or long and local or express), the corridor capacity, the propulsion technology (compressed natural gas or diesel), the type of payment (on-board or prepayment), as well as external circumstances such as the density of the corridor and the attractiveness of alternative modes influence the operational productivity of the system (defined by units of output per unit of operational input or the number of passengers boarding per kilometer of travel). Under these conditions, Bogotá and Jakarta exhibit the lowest operational productivities with only around five passenger boardings per bus-kilometer. Interestingly, the capital investment for TransJakarta averaged just USD1.35 million per km but the figure in Bogotá was the highest at USD8.2 million per km (Hidalgo and Graftieaux 2008). In spite of the achievements of these other forerunners, most BRT systems are for the most part modelled after the achievements of those in Bogotá, whose substantive infrastructure efficiently and quickly moves high volumes. Critics suggest that those same features which made Transmilenio so successful in Bogotá – wide busways, expensive stations and a sizeable fleet of articulated and bi-articulated buses – lead to rushed implementation, an overcrowded system, early deterioration of infrastructure, and glitches in the fare collection system, often associated with financial and institutional limitations, in cities of the global south (Hidalgo and Gutiérrez 2013).

In transportation studies, a policy model is sometimes referred to as a "policy package", or the combination of individual policy measures, aimed at addressing one or more policy goals created to increase the impacts of individual measures, minimize the negative side effects and expedite implementation (Filipe and Macario 2013; May 1995). The BRT model is an example of such a package formed to increase effectiveness, assure financing and other governmental support and accelerate public acceptance. A geographical approach considers the relational dynamics that direct the formation of such a package and in this case its reasons for distribution and acceptance. This chapter will now consider the emergence of the Bogotá model and its subsequent dissemination, to understand how policy models emerge and why they are applied across so many landscapes.

Forming the Bogotá Model of BRT

None of these previous iterations of BRT were as widely popularized as Bogotá's Transmilenio, which opened in 2000. When complete, Transmilenio will feature 388 km of dedicated busways (Mojica 2011). The first 14 km and 14 stations opened in December 2000 and the full 41 km of trunk was completed in June 2002, soon carrying 800,000 passengers per day. Construction on the 43 km of Phase 2 began in 2000 and opened in 2006 to carry 1.3 million riders per day. The third phase took another two years to complete opening another 28 km in 2009 (Cain et al. 2007). Transmilenio is just one aspect of a plan to decrease traffic congestion and alleviate air pollution in Bogotá. To complement the system, 300 km of paved cycleways were installed, 120 km of roadways are closed every Sunday, a license plate registration system known as Pico y Placa that bans 40 percent of the city's registered private vehicles during peak times (regulated by the last number on the license plate) was established, and Bogotá hosts the world's largest car-free day event each year. Transmilenio carries 1.3 million trips per day or 20 percent of the bus trips in the city on the 84-km network (Ardila 2007). Figure 3.1 illustrates how Transmilenio was the "gamechanger" because it was the first comprehensive BRT system built in a major metropolis to replace an unregulated and unruly transit industry. In addition to providing an example of the infrastructure of BRT – dedicated median lanes, iconic stations with high-floor platforms, off-board fare collection – Transmilenio importantly also demonstrated methods for incorporating existing transit operators, a major issue for urban transportation in South Africa. Its current propagation is driven by Bogotá's former Mayor, Enrique Penalosa, who served from 1998–2001 and again from 2016 to 2019, and his ties with the Institute for Transportation and Development Policy (ITDP), who together hype BRT as the most appropriate modern solution for those localities with limited financial and institutional capacity.

This discussion surfaces alongside an exploration of the international public transportation advocacy groups and research centers, providing evidence of BRT as best practice and providing capacity and financial support to those cities interested in replicating it. In addition to ITDP based in New York City,

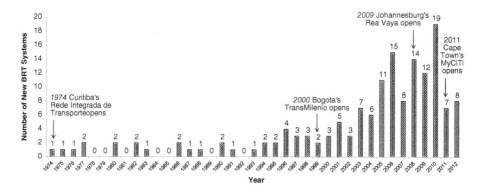

Figure 3.1 Number of BRT systems opening annually.

Embarq, the WRI Center for Sustainable Transport in Washington, DC, and the bus manufacturers, the Institute for Transportation Brazilian-based Marcopolo (now a joint venture with Tata Motors in Dharwad, India), Germany-based Mercedes Benz and the Volvo Research and Education Foundation (VREF) in Gothenburg, Sweden, promote BRT. A number of additional platforms dedicated to learning include the World Conference on Transportation Research managed by Embarq, the Sustainable Transport Congress in Mexico, and the TRB Committee on Developing Countries. The associations that provide a robust network of experts whose knowledge and expertise is necessary for implementation, will be explored further in the next chapter. The extent to which these agencies not only influence the decision to initiate a BRT system, but also the specifications of the height of the platform and the details of the bus fleet, is explored further in Chapter 4.

In October 1997, Enrique Penalosa was elected mayor of Bogotá with a promise to improve the bus system as a prelude to a metro system (Montezuma 2005). Penalosa's "mobility strategy" sought to prioritize public transportation over private vehicle use and to promote non-motorized transportation, in particular cycling. Bogotá had been experimenting with segregated facilities for buses since 1990, but Penalosa pursued a more intensive scheme to overcome the problems of the inadequate economic structure of public transportation, by introducing an unsubsidized system that also eliminated the violent competition between existing bus companies for passengers (Ardila 2007). Planning and construction began in 1998 with financial support from the World Bank and the national government of Colombia. The initial investment totaled USD240 million, 46 percent of which was financed through a fuel tax, 28 percent through local revenues, 6 percent by a credit from the World Bank, and 20 percent from the national government. International firms such as Steer Davies Gleave and McKinsey and Company were hired for their knowledgeable Brazilian consultants who had developed high capacity corridors in Curitiba, Porto Alegre and Sao Paulo. This work built upon previous studies including the Bogotá Master Plan and the Integrated Mass Transport System Study, both of which were devoted to the feasibility of a metro line. The planning team also visited systems in Curitiba, Quito and other South American cities to gain technical expertise and reduce opposition from transportation industry leaders (Hidalgo and Graftieaux 2008). This is a compelling instance when one city leveraged the experiences of another to execute an innovation. The structure and operations largely mimicked those of Curitiba and Quito but by learning from its peers, Bogotá was also able to experiment and expand on the model, in particular by demonstrating the financial sustainability of public transportation.

The main objective of Transmilenio was to create an affordable, comfortable, reliable transportation system as the impetus for a comprehensive urban redevelopment process. One study reveals that 83 percent of users perceive the increased speed as the main reason to use Transmilenio, while 9 percent of users have converted from commuting by private vehicle to the new transportation system. The fare of USD0.64 covers the operational costs of Transmilenio and the system operates without any subsidies, yet the performance is comparable to that of a rail system. Transmilenio is praised for its rapid realization: the project moved from a general idea to initial implementation in 35 months and the first 14 km and 14 stations were operational within four years. BRT was also particularly attractive to local policymakers because of its low cost – approximately one-third of a light rail transit project (Ardila 2007).

Problems with infrastructure (cracked concrete and rutted asphalt), operations (insufficient bus capacity to meet high demand), fare collection (long queues at stations) and user education (insufficient signage) did emerge over time (Hidalgo and Graftieaux 2008). In Bogotá, the level of service has declined in the initial corridors and several measures to counteract this are underway, such as the revisions of routes, construction of new connections among corridors and intermediate return points, the reduction of parallel routes of traditional buses, and the expansion of the bus fleet. Moreover, Transmilenio is still more expensive to build and complex to operate than conventional bus networks. Finally, the improvements introduced through the new system only eliminated the congestion and chaos along the 84 km of BRT network, and the more than 1,000 km of roadway used by the traditional system continue to dominate the market (Ardila 2007).

In spite of these criticisms, as soon as Penalosa's term ended, he began promoting the example of Transmilenio. In 2003, just months after Phase 1 of Bogotá's Transmilenio opened, Penalosa joined the Board of Directors at ITDP. Together they launched the "Building a New City Tour", the first BRT tour through South Africa, or as Paul Steely White, the executive director for Transport Alternatives and then the Africa Regional Director for ITDP called it, their "rock and roll tour of Africa" [41]. In the tour, Penalosa shared his experiences transforming

Bogotá to city officials in Dakar, Senegal, Cape Town and Tshwane and Accra, Ghana disclosing, "The people of Bogotá spent years hating their city ... Now, the people of Bogotá feel proud and have hope that their lives will improve. This is the story we are bringing to cities across the world" (White 2003). The tour was designed to help officials "build momentum for improving public transport" (White 2003) and enable local town planners to work with Penalosa and ITDP to devise strategies for improving their mobility. Penalosa's presentations therefore focused widely on transforming the urban realm. In recounting his motivations to build a more democratic and sustainable city, Penalosa emphasized improving the aesthetics of sidewalks to encourage walking and building new bikeways to inspire cycling as part of BRT projects [72]. By including these other elements into the BRT model, Penalosa and ITDP established it as a policy model for public transportation transformation, one that could be replicated in cities around the world. In South Africa, the model created a buzz around an otherwise politically obscure suggestion, thereby shifting the South African mindset (Kane 2003). White remembers, "People really rocked the transformative potential of giving major streets to buses, giving priority to buses over motorcars. ... What a powerful idea that instead of transportation infrastructure being used to fragment communities, it can be used to fuse communities". Although it took several years before BRT was adopted in South Africa, this tour inspired future thinking about urban development and transport planning. "People who were there and got excited about BRT... kept it alive and I think a lot of that did carry through to some of the projects that came to fruition in 2006" [41].

A deeper understanding of BRT reveals that the Bogotá model was not mimicked around the world, but rather a variation concentrating on the transformation of the colectivos emerged, which reflected the wider needs and opportunities in importing jurisdictions, and it became the exemplar model in South Africa precisely because it so aptly mutated. This suggests that policy models are polysemic and rather than duplicate elsewhere; they modify and mutate "often with different results, and occasionally resulting in a new, mutated policy approach for release back onto the circuit" (Prince 2012: 5). Although it seems to be a singular narrative of rapid mobility and perhaps even social justice, the introduction of BRT is rarely a straightforward, de-territorialized process but rather a highly politicized course of translation and mutation as BRT is plugged into different cities. Ong (2011: 14) points out that urban modeling takes "a set of normative and technical urban plans" and packages them as "a condensed set of desirable and achievable urban forms which sets a symbolic watermark of urban aspirations". In light of such affirmations, the next section demonstrates the variation in outcomes across South African localities, questioning the extent to which South African cities copied the Bogotá model as opposed to following a more adaptable process of interpreting and applying it to suit local needs and conditions.

Introducing BRT in South African Cities

Thirteen South African cities are at various stages in this process of adopting BRT and include Nelson Mandela Bay and Buffalo City in the Eastern Cape, Mangaung in the Free State, Johannesburg, Tshwane, Ekurhuleni in Gauteng, eThekwini and Msunduzi in KwaZulu-Natal, Polokwane in Limpopo, Mbombela in Mpumalanga, Rustenburg in North West Province, and Cape Town and George in the Western Cape. This study focuses on the adoption and implementation in the six cities whose projects are furthest developed – Cape Town, eThekwini, Johannesburg, Nelson Mandela Bay, Rustenburg, and Tshwane. Figure 3.2 provides a map of the cities which are building and planning BRT in South Africa.

The development of these BRT systems is depicted in Figure 3.3. Clearly, the Bogotá model was a critical influence on South Africa; however, the process through which BRT circulated among South African policy actors and was adopted locally, demands further study. The remainder of this section outlines the form and function of each of the cities and the methods for introducing BRT locally.

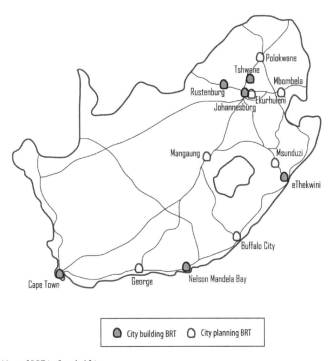

Figure 3.2 Map of BRT in South Africa.

This illustrative map of South African cities with BRT indicates the relative location of the 13 cities currently in various stages of planning and implementation. The six cities focused on in this study are highlighted for being advanced in their planning and implementation stages.

City	Johannesburg	Cape Town	Tshwane	Nelson Mandela Bay	Rustenburg	eThekwini
System	Rea Vaya	MyCiTi	A Re Yeng	Libhongolethu	Ya Rona	Go Durban!
Study Tour	August 2006; August 2007	November 2007; November 2008	July 2006 – Presentation by Lloyd Wright at Southern African Transport Conference August 2006 – National Department of Transport hosts Wright's visits to South African cities February 2007; September 2012 & November 2013	2007	November 2011 (South Africa); July 2014 (overseas)	2002
Council Support	November 2007	August 2008	May 2007; June 2011	May 2007	2009	May 2012
Construction Begins	October 2007	September 2008	August 2008	2008; and is stopped in 2010	June 2012	March 2014
Launch Date	August 2009	May 2011	July 2014	TBD	TBD	TBD

Figure 3.3 BRT adoption and implementation in South Africa.

Johannesburg's Rea Vaya

Johannesburg, with a population of 13.5 million (including 4.5 million in the City of Johannesburg, 3.2 million in Ekurhuleni Metropolitan Municipality and 2.9 million in the City of Tshwane) living across a highly urbanized, economically uneven, low-density landscape, is the largest city in South Africa (Statistics South Africa 2011). It is the main hub for the global trade in gold, diamond and platinum and as such it has the largest economy of any metropolitan region in sub-Saharan Africa (estimated at about 10 percent of sub-Saharan Africa's GDP), and it is the provincial capital of Gauteng, the wealthiest province in South Africa. While Johannesburg is not one of South Africa's three capital cities, it is the seat of the Constitution Court, which has the final word on the interpretation of the national constitution (South African Cities Network 2011).

Johannesburg is a poly-nucleated city with business and residential areas spread across the metropolitan area, extending into Ekurhuleni to the east and Tshwane to the north. Johannesburg covers 1,645 km^2, extending from Orange Farm and Lenasia in the southwest to Diepsloot and Kya Sand in the northeast, and from Roodeport in the West to Sandton in the east, with an average density of 3,000 persons per km^2. The city center is often cited as Africa's economic powerhouse because of the many commercial, financial, industrial, and mining headquarters located within the 6 km^2 business district, characterized by Victorian, Edwardian, Art Deco and modern architecture. The Carlton Centre and its 50 floors of shopping and businesses has been Africa's tallest building since it opened in 1971. Hillbrow, immediately north of the central business district, is one of the most densely populated residential areas in South Africa; while the residential and commercial areas in Randburg and Sandton are considerably more dispersed. Significant economic disparities are common in adjacent areas – for example Sandton, Johannesburg's most affluent commercial and residential district, is next door to Alexandra, one of the most deprived urban areas in the country. The city's residential composition varies accordingly, from extravagant gated mansions to informal settlements, driven by frequent population explosions mismatched with scarce housing stock. In addition, the city has been subjected to mounting crime rates, fueled by fear and scarring the urban fabric with security walls and armed private security.

Johannesburg's growth and development has been propelled by municipal trams, trolleybus and train services, as well as more recently, bus and minibus taxi services (for a more complete transportation history, see Chapter 6). The fact that Johannesburg is not near a large navigable body of water, has meant that road-based transportation has been the most important method of carrying people and goods. An orbital road made by connecting the N1 (which connects north to Tshwane and south to Vereeniging), the N12 (towards Ekurhuleni) and the N3 (to the east), surrounds the city center and the northern suburbs. In spite of it being up to 12 lanes wide at certain points, the ring road is frequently clogged with traffic, especially where the N3 Eastern Bypass and the R24 airport freeway intersect, which is said to be the busiest interchange in the Southern Hemisphere. Concurrent with the BRT project, the Province oversaw the Gauteng Freeway Improvement Project (GFIP), funded by national government at a cost of R23 billion, aimed at upgrading the main arteries to combat this congestion. A road tolling system, in which vehicles pay on a per-kilometer basis for the conveyance as they pass under one of 38 overhead gantries, was introduced in 2013 to offset the costs of the GFIP.

In August 2009, Johannesburg's Rea Vaya BRT system became the first full feature BRT on the African continent, promising to initiate a new era in South African public transportation. Construction on the 25.5 km and 30 stations of Phase 1A began in October 2007, just 16 months after political and technical leaders learned of the Bogotá model of BRT (See Figure 3.4). Rea Vaya, which translates from Sotho as "We are moving", includes three service types: trunk lines (for use only as busways), complementary routes (which collect passengers and then use the busway), and feeder routes (which collect passengers for transfer to trunk lines). The main trunk line (T1) runs from Thokoza Park to Ellis Park East with 20 stations along the way, and 3 complementary routes and 5 feeder routes. Stations are located roughly 500 m apart. The system utilizes 143 buses – 41 eighteen meter articulated and 102 thirteen meter rigid, double door. Rea Vaya was the only system in South Africa operational during the 2010 Football World Cup, moving 307,000 spectators. Each station cost R12 million per module and the cost per kilometer of busway was R36 million (not including bridges or provision for non-motorized transportation) for a total cost of R2.5 billion (see Appendix B on page 154) (City of Johannesburg Transportation Department 2011). Rea Vaya's success can be assessed through its ability to incorporate the minibus taxi industry into BRT operation – PioTrans, a consortium of former minibus taxi operators, currently manages Phase 1A – and its ability to continue building and operating subsequent phases. In February 2011, the 18 affected taxi associations formed PioTrans Pty with 313 shareholders and 585 taxis were withdrawn from operations.

Council approved Phase 1B in November 2008, and construction on the 18.5 km route between Noordgesig in Soweto and Parktown began in November 2010, with services beginning operation in October 2013. While construction of the 10 stations was complete by 2011, prolonged disagreement among the minibus taxi operators who have customarily provided transportation along the route stalled the project. In April 2014, Litsamaiso was formed from 10 affected taxi associations (74 percent of the company), Putco (22 percent), and Metrobus (4 percent) and 317 taxis were withdrawn. The system operates 134 buses – 41 eighteen meter articulated and 93 thirteen meter rigid, double door. Each station cost R12 million per module and the cost per kilometer of busway was R35 million (not including bridges or provision for non-motorized transportation) for a total cost of R1.2 billion. In July 2013, Rea Vaya started operating an automatic fare collection card in partnership with ABSA Bank, which led to a significant decline in ridership. Construction on subsequent lines has continued: once complete, Rea Vaya Phase 1A, 1B and 1C will include 122 km operated by 805

Figure 3.4 Fashion Square Rea Vaya station, Johannesburg.

Fashion Square Rea Vaya station is located on Mooi Street between Pritchard and Kerk Street in downtown Johannesburg. Fashion Square is a 26-block area in the eastern quadrant of the inner city, bordered by Market, Kerk, Von Wielligh, and End Street, specially earmarked for the development of a fashion industry. The iconic glass and steel architecture of the Rea Vaya station provides a beacon for the creation of a fashion capital.

buses. In total, the network will include 330 km of bus routes across Johannesburg with 85 percent of Johannesburg's population within 500 m of a Rea Vaya Trunk or Feeder service (City of Johannesburg Transportation Department 2011; World Bank, 2013).

Cape Town's MyCiTi

Cape Town, the second largest city in South Africa, is home to nearly four million residents. Located on the far western tip of the Western Cape Province, the city center is relatively compact, with a density of 1,100 persons per km^2 and a high concentration of office buildings and company headquarters; while the metropolitan area is significantly less dense stretching for more than 2,400 km^2. There is a strong mismatch between Cape Town's morphology and populated areas – the majority of the employment opportunities are in the city center and the inner suburbs, while the bulk of the population live in the southeast. Correspondingly, while the inner suburbs are predominately White and wealthy, the southeast is composed of makeshift shacks of rural migrants lining the borders of the city limits. This is evident in the range of city's population density – for example Cape Town's average density is 39 persons per hectare, but in the informal areas it can increase to as many as 150 persons per hectare and in the inner suburbs as low as 4 persons per hectare (South African Cities Network 2011). Of the 2.5 million trips made daily in Cape Town, 53 percent are made by private transportation, 38 percent by public transportation and 9 percent walking or cycling; and of the 38 percent made by public transportation, 18 percent are passenger rail, 12 percent minibus taxis and 8 percent bus (Mazaza 2017). In Cape Town, 95 percent of public transportation users are in the low- and middle-income brackets and spend up to 46 percent of their income on transportation (Jackson 2016).

Mitchells Plain is the largest formal township, originally built for Coloured residents in the 1970s and 1980s and now also home to significant numbers of backyard shacks because of housing shortages. Khayelitsha is another illustrious township developed for Africans in the 1980s and growing rapidly since, in spite of its poor access to employment opportunities or transportation. Atlantis is a third peripheral settlement that lacks sufficient transportation links to access the jobs in the city center. There is considerable scope for residential infill in the areas between the inner suburbs and the townships. The development of additional affordable housing would reduce transportation time and costs for commuters and increase the vitality of city center amenities. Promoting economic development in the southeast is a key focus for city and provincial authorities, with an overall mission to reduce the overall imbalance between city residents through expanded and efficient mobility solutions (South African Cities Network 2011).

Cape Town became similarly enchanted by BRT in January 2007[1] when Lloyd Wright and Ibrahim Seedat, Director of Public Transport Strategy at the National

Figure 3.5 Lagoon Beach MyCiTi station, Cape Town.
Lagoon Beach MyCiTi Station is the seventh of 12 stations along the T01 main route from Civic Centre to Table View. The station is situated in the median of the R27/Marine Drive in Milnerton.

Department of Transport, came to Cape Town to meet with Helen Zille, Mayor of Cape Town from 2006 to 2009. Lloyd presented the story of BRT and Bogotá's successful establishment of Transmilenio, and the Mayor responded with interest in launching BRT in Cape Town [54; 77]. There were subsequently two study tours – first a team of city officials and consultants went to Bogotá in November 2007 and then another group, which included minibus taxi operators went in November 2008 – to learn details of BRT construction and engineering, operations and maintenance as well as specifications for the rolling stock and non-motorized transportation [89] (see also, Wright 2007c). Construction began in September 2008 and MyCiTi Phase 1A opened in May 2011.

Like in Johannesburg, the objective for the MyCiTi IRT (Integrated Rapid Transit, as it is known in Cape Town) is to provide services within a 500 m radius for 75 percent of households. Services include trunk services, feeder services, trunk extension and supporting pedestrian and bicycle facilities (non-motorized transportation) with stations every 800 m. Phase 1A (sometimes referred to as the MyCiTi starter service) operates from outside the Civic Centre in the central business district, up the R27 to Table View (See Figure 3.5). Phase 1B, which expands the services within the West Coast corridor, from Du Noon to Century City via Montague Gardens, opened in November 2013. The total cost for Phase 1A including road works, stations and buses is expected to exceed R4 billion (total expenditure between 2006 and 2012 is R2.3 billion) and construction for

Phase 1B is expected to cost R710 million (City of Cape Town 2010; City of Cape Town MyCiTi Project Office 2012).

Among the city's major achievements is its intelligent transport system (ITS), which controls the bus services, ticketing and signage. Cape Town was the first South African city to employ an ITS including a smart card payment scheme, which improves operational management and enables intermodal transfer. The city's program also includes a comprehensive restructuring of the city's transportation services, which embraces the creation of a local transportation authority to act as the planning and contracting authority to manage and regulate MyCiTi operations, existing Golden Arrow bus services and aspects of the rail function (City of Cape Town Department of Transport, Roads and Stormwater 2012). A final achievement was the inclusion of the two former taxi associations into operating companies, TransPeninsula and Kidrogen in 2012. Two challenges with implementation were first the city's slow rollout, which meant that during the football world cup, MyCiTi operated shuttle services between the Civic Centre and the airport as well as a temporary loop around the downtown; and second, high operating costs make financing the system unfeasible. In 2011/12, MyCiTi earned R24 million from ticket sales but operational costs exceeded R30 million, leaving a shortfall of R6 million, which was covered by municipal subsidies (City of Cape Town MyCiTi Project Office 2012).

Tshwane's A Re Yeng

Tshwane (also known by its historical name, Pretoria), the nation's capital city and the second largest city in Gauteng, is home to nearly three million people. The municipal boundaries are substantial, stretching from Centurion in the south to Soshanguve in the north for a total area of 6,298 km^2. Much of the urban economy is situated around the Union Buildings (the official seat of the South African government) and the offices of the president, as well as hundreds of international residencies and embassies and their associated donor affiliates in Arcadia and Hatfield. While the city center hosts a number of banking headquarters, corporate offices, small business, and shops in addition to central government offices including the Department of Transport and the National Treasury, Tshwane's main population centers – such as Soshanguve, formerly part of the Bophuthatswana homeland and only incorporated in 2000 – are located some 25 km beyond these employment opportunities (see Figure 3.6). It is thus more apt to characterize Tshwane by its highly uneven population distribution than as low-density urban sprawl. Tshwane, like all South African cities, has suffered from chronic underinvestment in public transportation. A study of average trip length for car users found that Tshwane residents travel twice as long as residents of London, Singapore and Tokyo, and about three times as long on public transportation. This can be explained by the low density of the city and displaced urbanization as a result of apartheid spatial planning (South African Cities Network 2011).

Tshwane was also eager to apply BRT locally and in February/March 2007, 15 representatives from the transportation and planning departments as well as members of the mayoral committee and the bus and minibus taxi industry, visited Bogotá and Paris and Rouen, France. The city organized and paid for the tour and the Department of Transport was not involved at that point. Following the study tour, a report was submitted to council detailing its' outcomes, and in May 2007, Lloyd Wright and his consultancy Viva Cities completed the

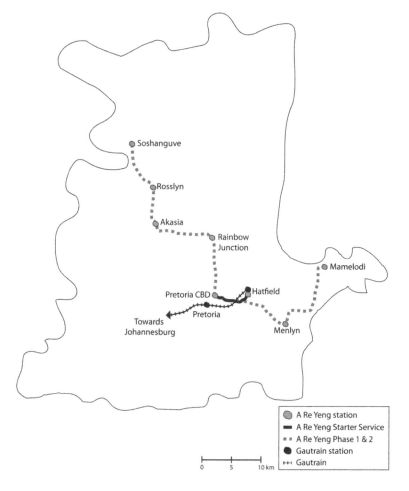

Figure 3.6 Map of A Re Yeng, Tshwane.
This illustrative map of Tshwane's A Re Yeng system indicates the relative location of the proposed starter service and Phases 1 and 2.

Tshwane Rapid Transit: Implementation Framework (Wright 2007d), which was approved by cabinet. Planning began in August 2008 but construction on A Re Yeng, meaning, "Let's go" in Sotho, was stopped in 2010 due to conflict between national government and the municipality regarding the alignment to Soshanguve.[2] National government was concerned that the BRT route duplicated the rail services along the M17 and National Treasury intervened withdrawing its financial support (R2.5 billion) for the infrastructure. This is the only instance when national government blatantly interfered in the realization of BRT [39]. Continuity has also been a major challenge for this project. Between 2007 and 2010, Tshwane had three mayoral members and four project leaders. Without "political buy-in" and "constant top management support", Tshwane's BRT "never got off the ground" [42].

With the support of national government, in June 2011 the city again began to build a BRT in Tshwane. In November 2011, the city set up a project office and submitted a revised strategy for A Re Yeng to Tshwane government for approval. Construction on a revised 7-km starter service from Paul Kruger Street in the center of Pretoria to Hatfield CBD via Sunnyside, that also connects riders with the Hatfield Gautrain station, began in July 2012 and started operating in July 2014 (See Figure 3.7). Once completed, the system is expected to cover 70 km. Tshwane is the first South African city to construct low-floor platforms (340 mm), claiming that their stations are an improvement over Rea Vaya and MyCiTi

Figure 3.7 Hatfield A Re Yeng station, Tshwane.

Hatfield Station is the first A Re Yeng station to be completed in April 2014 along Line 1 from the CBD to Hatfield. This station is situated in the median lane of Arcadia Road in Tshwane. The station's "retro tram concept" is evocative of the city's old tramway lines, which were removed in 1939.

because they are more accessible. Tshwane is also pioneering a new concept for acquiring the 171 buses (46 articulated, 125 rigid): rather than the city owning the buses, they facilitated a loan for the to-be-formed bus-operating company to purchase the buses rather than selling them the buses after five years, as in Cape Town and Johannesburg. At R50 million per kilometer and R10 million per station, A Re Yeng is also the most expensive system in South Africa. The goal for A Re Yeng is to provide improved services for 85 percent of Tshwane's population within 500 m of their homes [42].

Rustenburg's Yarona

Rustenburg, with just half a million residents, is the smallest municipality to adopt and implement BRT. It is the largest city in the North West province and capital of the Royal Bafokeng Nation. Rustenburg is one of the fastest growing cities in South Africa because it is home to two of the world's largest platinum mines responsible for refining 70 percent of the world's platinum. This impressive economic vitality is accompanied by investment in social and spatial integration.

Rustenburg's Yarona, meaning "It is ours" in Setswana, will include 34 km of dedicated infrastructure and 32 closed stations, running along 6 main routes into the central business district and connecting to a considerably larger network of feeder buses (Froschauer 2012). Services include 2 trunk lines (5 km along the R510 to Kanana and 7.5 km along the R565 to Phokeng to the city center), one direct line (on the R24) and 9 feeder lines, for a total of 240 stops across the network (See Figure 3.8). Yarona will operate 233 buses in Phases 1 and 2, covering 18 bus routes and is expected to carry 180,000 passengers daily. Each station is expected to cost R8 million and the total cost of the project is R3 billion, including R2.4 billion for infrastructure and another R680 million for buses. Construction began in June 2012 and the system launched in 2021 [75].

Rustenburg is often condemned for building a BRT with critics claiming the city is too small to sustain the expensive system, while others claim that Yarona perhaps has more potential to influence that city's social and spatial integration than larger cities, whose form is less likely to change through the intervention. Whereas a BRT in a larger city may take 20 years to complete, Rustenburg's entire network is much smaller and construction was expected to take just five years [75]. In addition to being the second South African city to construct low-floor platforms, several other characteristics make Yarona unique. The vast majority of households (84 percent) in Rustenburg do not have a car available for private use – this figure is significantly higher than most other South African cities. Many Rustenburg residents walk to work because they cannot afford any motorized forms of transportation. It is therefore important for Yarona to reach at least 85 percent of Rustenburg residents within 1 km of their homes. Moreover, because Rustenburg is a local municipality,[3] it does not have a transportation or

Figure 3.8 Yarona station platform, Rustenburg.
This nearly complete platform along the R510 outside Thlbane is an example of the 32 low-floor stations expected in Rustenburg. Notice how the roadway is elevated to prevent interaction with ordinary traffic.

planning office; in 2009, the mayor appointed Namela, a transportation-planning consultancy directed by Pauline Froschauer to manage the project. Since 2011, construction and planning for Yarona has been based out of a new BRT project office, which maintains the smallest staff of any city with just 17 employees, many of whom joined through the Transport Rustenburg Incubation Program (TRIP), the first of its kind for a BRT project. Finally, in November 2011, representatives from Rustenburg's taxi industry visited operational systems in Cape Town and Johannesburg, instead of heading directly to Bogotá. When they returned, the 11 affected taxi associations in Phase 1 have formed the Taxi Negotiation Forum to liaise with the City throughout the transformation. In July 2014, midway through the transition process, the group travelled overseas to see BRT systems in South America and meet with operators [96].

Nelson Mandela Bay's Libhongolethu

With a population of just over one million inhabitants, Nelson Mandela Bay (Port Elizabeth), in the Eastern Cape province, is one of South Africa's eight Metropolitan municipalities (or a Category A locality according to the Municipal Systems Act of 2000). The city is aptly named after the country's first democratically elected president. The municipality has a sprawling city form – the city center has a relatively low residential density and is considerably disconnected from larger, denser residential areas of Motherwell, New Brighton and

Kwazakhele. Efficient transportation connections are vital to link township residents in the north with the jobs and facilities in the center of the city (South African Cities Network 2011).

Nelson Mandela Bay was the first city charmed by the idea of transforming its public transportation network through BRT, and the second city to approve a BRT plan in council and visit Bogotá in 2007. The plan for Libhongolethu, meaning "Our pride" in Xhosa, called for five routes – the Khulani Corridor, Kwazakhele, Motherwell and New Brighton to the city center via Korsten – the first of which was the 10 km along the Khulani Corridor to the center of Port Elizabeth. Guiding the project was the motivation to provide an efficient, safe, affordable, sustainable, and accessible multimodal public transportation system, which supports social and economic development to ensure optimal mobility and improved quality of life for the residents and users of the transportation system in the metropolitan area (Nelson Mandela Bay Steering Committee 2010). In 2008, construction began but then halted in 2010 when the project was besieged by politics and poor planning, as well as weak municipal leadership, which failed to reconcile with the minibus taxi industry (IOL News 2010). R1.3 billion was allocated before the stoppage in 2010. In July 2010, the municipality purchased 25 buses, which sat idle for three years. On 29 January 2013, the municipality handed over 19 buses, bought during the 2010 Football World Cup, to TransBay, a consortium which ran for 12 months, to operate a pilot program. In November 2013, the pilot ended and it remains unclear, if or how, the municipality will continue to realize its BRT system [95].

eThekwini's Go Durban!

eThekwini (also known by its historical name, Durban), the largest metropolitan area in the province of KwaZulu-Natal, one of the main the business seaports in Africa, and the second most important manufacturing hub after Johannesburg, is home to 3.5 million residents. Its tropical climate and sandy beaches make it a center for South African tourism. eThekwini is connected by road, rail and air to other cities in South Africa and the wider world. The municipality is considerably larger than other South African cities occupying nearly 3,000 km^2 with a lower density of approximately 1,500 persons per km^2. eThekwini thus has a more fragmented spatial form with three distinct residential zones – the city center and inner western suburbs, the northern townships including KwaMashu, Inanda and Lusaka, and the south-western township around Umlazi – with the remainder of the metropolitan area consisting of low-density suburban sprawl and traditional rural areas. This divided morphology is also reflective of the regional topography of undulating hills and valleys, which thwart efforts to strengthen transportation connections between the city center and main residential settlements. There is scope for residential development around the city center and inner suburbs, which would strengthen economic opportunities in the city center (South African Cities Network 2011).

eThekwini has been one of the slowest cities to develop a BRT project. Representatives from the eThekwini Transport Authority, one of the only such establishments in South Africa, were among the first South Africans to visit Bogotá, participating in an early study tour in 2002. While in Bogotá, officials noticed both similarities, especially in terms of their unregulated taxis, as well as differences, including the variations in population, density, employment and income between Bogotá and South African cities, and returned with reservations regarding the applicability of Bogotá's BRT system in eThekwini. This led the city to postpone introducing the concept locally and instead focus on improvements to existing rail systems [47]. In May 2012, in spite of these earlier concerns, the municipality approved plans to proceed with the first three lines of the Go Durban BRT system, slated to be completed in 2030. In August/September 2013, three representatives from the eThekwini Transport Authority visited Buenos Aires, Bogotá, Cali, and Sao Paulo to investigate the IT systems that support bus operations including bus management, traffic management and passenger information systems, and expect to integrate this learning into the new system. Officials at the eThekwini Transport Authority rationalize this delay as part of a comprehensive wall-to-wall planning process through which it created a network of nine trunk corridors, supported by feeder and complementary services that will integrate bus, rail and taxi systems into a seamless, professional service. Another cause for delay was litigation with the Early Morning Market traders at Warwick Avenue, who were to be relocated as part of project but will now be incorporated into the new terminal [49].

Construction on Phase 1, which includes C1 (Bridge City to the city center), C2 (Bridge City to Umlazi by rail), C3 (Pinetown to Bridge City) and C9 (Bridge City to Umhlanga), began in mid-2014 with a completion date of 2027. The first 57 km of trunk corridor, 181 km of feeder routes, 42 stations and 297 stops on the feeder routes, 4 terminal stations, and 16 depots of Phase 1 will accommodate 65 percent of the city's public transportation demand (25 percent on C1, C3 and C9 and a further 40 percent on C2 rail corridor). Only the C3 corridor will include a dedicated median lane as well as low-floor platforms and closed preboarding fare collection stations; the other stations will utilize a less infrastructurally intensive system. Phase 1 will cost R10 billion and the total price for Go Durban is expected to reach R22 billion, of which the city will pay R20 billion while the Passenger Rail Agency of South Africa (PRASA, the parastatal responsible for rail across South Africa) will spend R2 billion building rail networks and upgrading existing lines.

The above profiles provide the background of the six BRT projects in South Africa. While there are a number of different approaches to BRT in South Africa, the central thread is a mission to improve the public transportation options for urban commuters. Although planning has been based on the operational system in Bogotá, these plans demonstrate a uniquely South African interpretation of BRT. On the Cape Town leg of the 2013 Rustenburg study tour, Jeremy Cronin, then-Deputy Minister for Transport remarked, "The importance of the engagements such as these, is that they begin to craft a South African model in as far as public transportation transformation is concerned, of course based on our shared practical experiences" (2013). In South Africa, BRT offers an opportunity to integrate current bus and taxi operators into a single system, which enables operating subsidies to be contained and an end to the discrimination in subsidy arrangements between buses and taxis. Operators are paid for running set routes irrespective of the number of passengers transported, and thus are not expected to chase passengers in order to maintain revenues. The next section considers the physical characteristics of BRT in South Africa and evaluates these material facets in comparison to their counterparts elsewhere.

A South African Interpretation of BRT

No two BRT systems are exactly alike. Each has unique characteristics tailored to successfully operate in its specific geopolitical context. ITDP's BRT Planning Guide (Wright 2007a), however, suggests cities realize as many as possible of the following features: an integrated network, comfortable and secure stations, level access between the platforms and buses, preboarding fare collection and integration between routes, restricted entry to the system, distinctive marketing features, low-emission vehicles, centralized control center and bus management technologies, special provision for physically disadvantaged groups, and clear route maps and schedules. In 2010, ITDP published the BRT Standard, a design guide and rating system that establishes a common definition for BRT and develops a scoring mechanism to evaluate BRT corridors and recognize those with superior design and operations. The Standard certifies the systems as basic, bronze, silver or gold-rating system and those which fail to meet the minimum standard for basic are classified by ITDP as "not BRT".

Interestingly, each South African city uses a different common name for BRT: Cape Town calls its Integrated Rapid Transit; in Tshwane and Rustenburg, it is the Tshwane Rapid Transit and Rustenburg Rapid Transit, respectively; eThekwini uses the term, Integrated Rapid Public Transport Network; and in Nelson Mandela Bay, it is Integrated Public Transport System; only Johannesburg calls it BRT (see Appendix B on page 154). This suggests a profoundly different understanding of the infrastructure and intention for the system. One could say that Johannesburg's Rea Vaya is more closely modelled on Transmilenio, while Cape Town tried to integrate the new transit system with existing rail and minibus taxi networks with the overall intention of forming a municipally controlled and integrated transportation authority. The use of the "network" rather than "transit" suggests a more comprehensive approach. Noticeably absent from all but Johannesburg's terminology is the focus on the bus. This is an important point as I unravel the South African version of BRT.

Both Cape Town and Johannesburg's BRT systems complied with the standards created by Transmilenio and circulated globally but when introduced in South Africa, these material features were interpreted and applied within the local spatial-political context. Figure 3.9 provides an overview of the specifications of the planned and operational systems in South Africa, while Appendix B on page 154 provides additional details of the other South African cities. My arguments regarding the material and immaterial components of BRT are rooted in the way in which South African cities developed their own version of the technical and procedural elements of the BRT model.

Features of BRT Systems in Cape Town and Johannesburg		
	MyCiTi Phase 1A, Cape Town	Rea Vaya Phase 1A, Johannesburg
Operational since	May 2011	August 2009
Services operating	2 trunk services; 9 feeder services	1 trunk route; 3 complementary routes; 5 feeder services
Management agency	City of Cape Town, Department of Transport	City Transport Department (contracted services unit)
Operating company	TransPeninsula and Kidrogen	PioTrans (Pty)
Automatic fare card launch date	February 2012	July 2013
Raise lane delineator	Kassel curb	Concrete rumble blocks
Asphalt	Red asphalt included in original design	Red-painted line through the centre of the roadway retrofitted in February 2012
Height of the station	High floor (940 mm)	High floor (940 mm)
Number of buses	267 buses in operation	143 buses in operation (41 articulated, 103 rigid)
Number of stations	17	30
Network length (in km)	16	25.5
Cost per km	R40 million	R36 million (excluding stations)
Cost per station	R12 million per module	R4 million per station
Cost per trip	R25 (myconnect card), R6.80 (0–5 km), R7.90 (5–10 km), R9.60 (10–20 km), R12.70 (20–30 km), R14.30 (30–60 km), R21.10 (>60 km)	R25 (smart card), R11.50 (25–35 km), R12.50 (35 + km), R13 (occasional user one trip), R25 (occasional user two trip)
Weekday patronage	30,000 (expected patronage: 100,000)	45,000 (71,000 maximum forecasted)

Figure 3.9 Features of BRT Systems in Cape Town and Johannesburg.

In South Africa, BRT had quite a range of interpretations: several interview respondents explained the technical aspects of the system, while others presented the way in which BRT resonated with ongoing planning and practices in South Africa. Most South Africans were proponents of BRT, extolling the efficiency of the system and its transformative effects on both the economic and social life of the city. Many of these same respondents simultaneously expressed concerns over the cost of building and operating the system, and its ability to reach the marginalized sections of the city. People described BRT as a "multifaceted urban intervention" [8], "infrastructure project" [39; 54], "consultant-driven process" [83], "politically-driven" [22], "radical transformation" [94], "transport solution" [92] and "expensive" [20]. Architects described its physical features [34] and development specialists described it as "not one solution but a range of solutions" [32]. I was told the importance of the acronym BRT which takes the bus out of the equation, and therefore some cities designed streamlined vehicles and stations that more closely resemble train cars [72], while still others told me of the importance of BRT as a tool for "social restructuring" [70].

Respondents replied with these various answers because there is no precise definition for BRT. BRT is the latest "buzz word" (Jarzab et al. 2002) among transportation planners, promoting it as the economic and practical solution to urban congestion and immobility. It is both a technology for moving people from A to B as well as a transformative tool for urban densification. "There is not one model of BRT that can be applied across all cities" according to Ajay Kumar, Senior Transport Economist at the World Bank, rather BRT spread through South Africa because it had multiple meanings to implementers. Rehana Moosajee, Member of the Mayoral Committee (MMC) for Transport in Johannesburg from 2006 to 2013 and a pioneer supporter of BRT, defends her decision to institute BRT, reasoning that it fit well with the city's political vision for economic, environmental and spatial transformation, and was relatively cheaper than other modes of transportation – so it had the additional impact of being possible [22; 80]. "It is not so theoretical, out-there", Moosajee inferred, "[there] is the manual ... [there] are these principles and if you think it works for your city then introduce it in your city. That is what I like best about his approach" [22]. For South Africans, BRT spoke to many principles of "social mobility, governance, poverty alleviation" [7], conjuring up images of an accessible, compact, integrated city, values which lacked a codified form. Graeme Gotz, Director of Research at the Gauteng City Region Observatory (GCRO), was one of the first interview respondents to question the idea that BRT as one thing, suggesting that "Some projects like BRT speak to many projects, not just one project. BRT was critical because it spoke to a number of those principles" [31].

In addition, Maddie Mazaza, Director of Transport in Cape Town, rationalized integrating her own learning with the corporeal aspects of BRT. "For me, learning from other case studies is about learning from the principles and not replicating what they did", she explained. "If they say it didn't work, you need to see that in their situation maybe it was different because of funding, land-use or the environment" [70]. This almost transcendental quality, which makes BRT seem relevant in spite of various local policy aims and conditions that might otherwise make it inappropriate, is what I will argue enables its translation as it moves from place to place around the world.

About the Station Platform

Station platforms are the most conspicuous feature of the BRT system. They were introduced by Curitiba in 1991 and now form a key component of ITDP's BRT Standard, as part of an assortment of BRT characteristics duplicated around the world (Rabinovitch and Hoehn 1995). BRT stations are designed in modular units to accommodate the construction of numerous station configurations cheaply and easily. In Johannesburg, there are two standard platform modular widths – 4.8 m and 3.3 m [34]. Platform heights range from low floor (220 mm) – now considered the gold standard for easier access for step-free users – to high floor (650–1,100 mm) – more common in those systems designed in the mid-2000s, because the raised platform on the buses could hold a higher capacity of passengers and was also cheaper to purchase (Venter et al. 2004). BRT systems seek to minimize the gap between platform edge and bus floor, either by means of an on-board bridge or a positioning system, which helps ensure that the bus is stationed immediately adjacent to the platform at stops, eliminating the need for a bridge (Manning 2007, 2009). Alignment markets, Kassel curbs and boarding bridges help facilitate a standard gap of less than 5 cm (2 in.). The notion of level boarding, which enables easier and faster entry/exit to/from the bus, is one of the key methods for reducing docking times at stations, and necessary for both customer safety and comfort.

The decision to introduce high/low-floor platforms in South Africa was informed by concerns for the safety and comfort of riders, as well as the compliance of the existing transit providers. Whereas Cape Town and Johannesburg opted for a 900 mm high-floor platform (see Figure 3.10) to prevent existing transit vehicles from stopping at the stations, the systems in Rustenburg and Tshwane use a low-floor platform of 340 mm, rationalizing that these lower platforms were cheaper to build, easier to access and thereby justifying the higher costs for the bus. Compelling experiences from Bogotá led South African localities to duplicate the specific height of the platforms in Transmilenio, which were designed in partnership with local operators who agreed to the physical incompatibility with the existing system (Pineda 2010). While it may seem like a simple engineering component, the trust built from developing the height of the platforms alongside the informal operators established a confidence in Bogotá that drove the success of Transmilenio. In South Africa, however, this prescribed height, although established, had no local salience because minibus taxi operators were not similarly consulted regarding those specifications. Cape Town and

Figure 3.10   Rea Vaya high-floor station, Johannesburg.
"The City's BRT stations are designed to be naturally ventilated and insulated from heat and cold; they are lightweight, and visually and aesthetically appealing", Mayor Amos Masondo told the audience at the grand opening of a Rea Vaya station in 2008. "Gone are the days of the sad, graffiti-marked little bus shelter, where wind, rain and sun treated commuters with equal disdain. The futuristic design of the Rea Vaya station is made of steel, glass and concrete and is solid, safe and roofed" (Masondo 2008).

Johannesburg, because they were first to design their systems, were more heavily informed by the Bogotá model. Later adopters like Rustenburg and Tshwane were more inclined to learn from experiences elsewhere and follow the universal trend towards low-floor platforms.

About the Bus

Another feature of the BRT model is the bus, which is typically designed to carry high volumes of standing passengers. The South African model of BRT however includes a major social component by providing all passengers with a seat. For Rehana Moosajee, the "dignity of commuters" was of the utmost importance [22]. Recall that because of South Africa's dispersed spatial form, many South African commuters have exceptionally long, arduous commutes. "Part of the problem with existing public transport in South Africa", explained Lael Bethlehem, Chief Executive of the Johannesburg Development Agency (JDA) from 2005 to 2010, "is that people are treated like sardines". On BRT, "everyone has an individual seat and a seatbelt" [36]. For example, Johannesburg's Rea Vaya operates 143 buses – 41 articulated and 103 rigid (see Figure 3.11). The larger articulated bus carries 117 passengers – 61 standing and 56 seated. Cape Town's articulated buses carry 53 seated and 77 standing, the smaller 12 meter rigid buses carry 43 seated and 44 standing, and the 9 meter minibuses carry 25 seated and 25

Figure 3.11 Rea Vaya bus, Johannesburg.
Johannesburg's Rea Vaya operates 143 buses – 141 articulated and 103 rigid buses – carrying between 50 and 130 passengers. Because the look of the bus is particularly important for the overall brand of BRT, the buses were designed in partnership with local architects and designers.

standing. The buses in Tshwane are even smaller – the 18 meter articulated buses carry 90 passengers with 44 seats and the 12 meter are suited for 60 passengers. South African BRT buses are smaller, with more seats and less standing room than their counterparts in Bogotá [22]. It was "a policy decision that for us public transport is about restoring dignity and that is quite different from what happened in Bogotá" rationalized Bethlehem [36]. However, the viability of BRT is based on high volumes of people moving quickly across the city. The Bogotá buses are designed to hold considerably more passengers (as many as 270 passengers) than those in South Africa (which carry between 50 and 130 passengers). As a result, even when the system reaches its expected ridership, patronage will still be lower than initially estimated while operational costs will be higher.

About the Bus Lane

A dedicated right-of-way is another feature of the form and function of the BRT network. It facilitates the rapid and unimpeded movement of the buses. Dedicated lanes matter the most in heavily congested areas, where it is harder to take a lane away from mixed traffic to dedicate it as a busway. Physical design is critical to the self-enforcement of the right-of-way. Lane segregation is enforced to varying degrees of permeability through raised lane delineators, electronic bollards, car traps, colorized pavement, and camera enforcement. In some cities, the bus stations themselves can act as a barrier. Some permeability is generally advised, as buses occasionally break down and block the busway or otherwise need to leave the corridor. Bogotá and New York rely on red-painted asphalt but no raised delineators, while Ahmedabad, Quito and Mexico City use yellow concrete delineators but no indication through the center of the road (Hidalgo and Graftieaux 2008).

Johannesburg initially delineated the busway from ordinary traffic with concrete yellow rumble blocks. In February 2012, engineers retrofitted a red-painted line through the center of the roadway to prevent ordinary traffic and minibus taxis from entering the BRT lanes (see Figure 3.12). In just a few years since opening, the rumble blocks have deteriorated significantly, and in Phase 1C, Rea Vaya lanes will be outlined by yellow plastic rumble blocks, which are expected to last longer [5]. In Johannesburg, these demarcations have proven unable to keep the BRT lanes free of non-BRT traffic. Respondents cited problems keeping taxis and passenger vehicles from driving in the BRT lanes despite attempts to fine drivers. This significantly reduces the efficiency of the system and downgrades the brand [29; 88; 89; 92; 95]. Moosajee in Johannesburg told me with evident frustration that even politicians are driving in the BRT lane presuming that they are exempt from regulations limiting it to BRT buses [22]. Whereas some argue that strict adherence to the BRT lanes is necessary for system efficiency [81], others suggest that perhaps these BRT lanes are the wrong technology for South Africa and the system would be cheaper to build and easier to operate if buses moved through mixed traffic [75].

Cape Town experimented with the Kassel curb, a concave section usually used to service low-floor buses first introduced in the German city of its name (see Figure 3.13). The

Figure 3.12 Rea Vaya bus lane, Johannesburg.

Johannesburg's Rea Vaya Phase 1B takes passengers along Joubert Street from the city center, past Park Station through Braamfontein. Pictured here is the busway on Rissik Street traveling north to the Johannesburg Metro Building.

Figure 3.13 MyCiTi bus lane, Cape Town.

The City of Cape Town placed MyCiTi along the median of the R27. Cape Town was the first city to use red asphalt to delineate the lane, a feature later mimicked in Johannesburg.

rounded section allows the driver to align the tire of the bus with the station to enable a closer connection with the station. Cape Town also laid red asphalt for an additional R2.5 million, but it has been so successful that Johannesburg initially considered retrofitting the entire lanes but later compromised with the red-painted line. "A simple feature like selecting the color of the road to be red", claims Eddie Chinnappen, then-Executive Director of Transport in Cape Town, "can become the hallmark feature of your city's system" [89].

Interestingly, both Nelson Mandela Bay and Tshwane copied the Kassel curb from Cape Town and the red-painted line through the center of the roadway, revealing the influence of both cities in its planning and construction. eThekwini also duplicated the Kassel curb and the red asphalt used in Cape Town. Rustenburg, however, opted against either, instead constructing a 200 mm raised roadway which should prevent ordinary traffic from entering the lane.

About the Route

Decisions regarding the routing of the BRT network are said to be among the most important engineering features, and concurrently reflect deeply embedded sociopolitical concerns, in this case for improved public transportation provision. Thinking in terms of policy mobilities, however, these decisions are indicative of the materiality of the BRT model, which has strong visual and corporeal features but still vary when applied in different contexts. Cape Town's decision to service the largely White, middle-class West Coast suburbs was heavily criticized but planners justified their decision claiming that this was the only area of the city not served by rail, it would reduce traffic congestion and, most importantly, it had the fewest number of taxi associations, thereby reducing the tension of formalization (see Figure 3.14). Whereas Capetonians elected to build its starter service up the

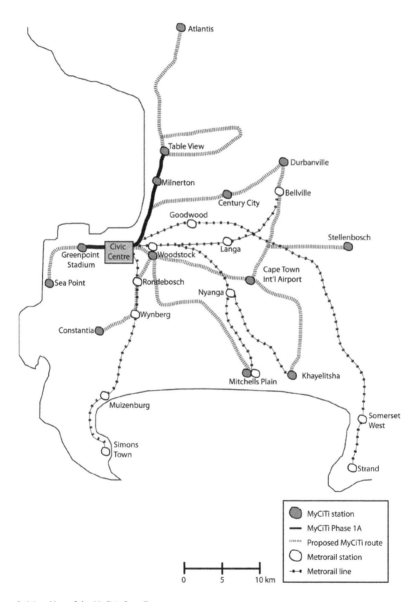

Figure 3.14 Map of the MyCiti, Cape Town.
This illustrative map of Cape Town's MyCiTi system indicates the relative location of the currently operating Phase 1A and the proposed future phases. It also indicates in black the existing Metrorail line.

TRANSLATING BRT TO SOUTH AFRICA   57

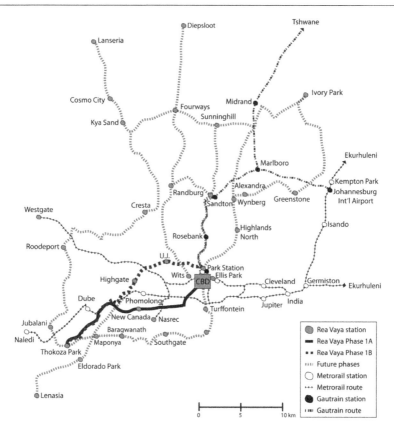

Figure 3.15   Map of Rea Vaya, Johannesburg.
This illustrative map of Johannesburg's Rea Vaya system indicates the relative location of the currently operating Phase 1A and Phase 1B, as well as the proposed location of Phases 1C and 2. It also indicates in blue the route of the currently operating Gautrain, a high-speed rail service connecting Johannesburg and Tshwane, which opened in 2010, and in black the existing Metrorail line

West Coast, in an area not previously served by other forms of public transportation, Joburgers placed the first BRT route between Soweto and the city center, servicing the poorest people. Johannesburg's alignment was criticized for catering towards the ephemeral needs of the 2010 Football World Cup by positioning the route's terminus at Coca-Cola Park in Bertrams and its divergence through the FNB Stadium in Soweto (see Figure 3.15).

Both networks differ considerably from their exemplar in Bogotá because unlike Transmilenio, which moves through a dense, urban context, South African BRT networks traverse across sparsely populated areas and thus passengers typically only board at the beginning and alight at the terminus of the line. This long haul style route contributes to low ridership figures, high operational costs and reduced economic impact on the area surrounding stations and supports arguments that BRT is perhaps inappropriate for the South African city. It is therefore important to associate these decisions with the local political landscape, which has focused, perhaps too heavily, on incorporating the minibus taxi industry rather than considerations for the physical landscape. Both cities therefore selected routes which either had the fewest taxi associations as in the case of Cape Town or with whom the city had already built a relationship as in Johannesburg. As it relates to this research, these examples demonstrate that BRT support therefore differs from place to place both within South Africa and globally, and the outcome is affected by localization in spite of close mimicking.

Physical features (i.e. busways, platforms and buses) suggest a certain standardization of the BRT model and while they may seem easy to replicate (i.e. build platforms that are 900 mm, demarcate lanes using expensive engineering specifications and utilize large-capacity buses) in fact, BRT is not an easily reproducible technology. The BRT model is not universal because its application is contingent on local circumstances, including attitude towards crowded buses, lane demarcations and so on. While residents of one city may respect a red-painted line instructing drivers not to enter the BRT lane, in another city drivers may need larger bollards to prevent them from driving in the BRT lane. The South African version of BRT therefore differs from its counterparts elsewhere – whereas the Bogotá BRT model was favored for being cost-efficient and technically feasible, in South Africa, BRT projects were guided by a concern for restoring human dignity by introducing more choices for public transportation users. This overarching social justice bent led to considerable differences across the projects. Lael Bethlehem recalls, "We followed the guidelines and rules for building BRT but the way it came out is slightly different" [36]. In South Africa, BRT was favored because it "resonated with our approach to transit-oriented development, which here is more of a concept than just an engineering dimension", concluded Rashid Seedat, Head of Gauteng Planning Commission. For ideas from elsewhere to be employed locally, he added, it has to be "something that we are already doing anyway" [40]. In order for BRT to seem as if it belonged in South Africa, it had to mutate into whatever each city needed: in Cape Town, BRT was perhaps guided by environmental concerns; while in Johannesburg, it was attractive for its ability to reach the marginalized sections of the city; and in eThekwini, BRT was favored when it combined with ongoing efforts to reform the rail system. BRT is such a potent model – but also fails – because it is not a universal technology, but rather a process through which its meaning changes into whatever best fits the importing locality.

BRT and Taxi Transformation

The adoption of BRT in South Africa offers an occasion to investigate the various components of the BRT model. In this previous section, I demonstrated how BRT serves as both a technology for moving people as well as a transformative tool for urban densification. BRT advocates both within South Africa and globally argue that building stations and busways are the "soft issues" which are "soft in nature but require a lot of engineering capacity". The "big issues" by contrast are "much more complicated because they involve people" [33], referring to the challenge of integrating the informal transit operators into the formal BRT network. This section focuses on the transformation of the informal taxi industry, a key feature of the current popularity of BRT.

In South Africa, where a poorly regulated, aggressive and often violent minibus taxi system competes for passengers, often at the expense of riders and road safety, BRT is a tool through which government can intervene in the industry. The transformation of an informal transportation service into a formal industry, operating within the quality frameworks required by government, is not as easily replicated as the aforementioned material features of the model. There is distinct local logic to the taxi industry and its historical and ongoing resistance to government interventions, which make its transformation challenging, in particular because there is no system of trust between the various stakeholders. This last empirical section of this chapter considers the growth of the minibus taxi industry, the subsequent post-apartheid legislation to regulate and manage it, and the process by which the incumbent operators are incorporated into the new BRT system.

The South African Taxi Industry

The informal transport industry emerged in South Africa in the 1970s as a self-made, Black entrepreneurial response to the failures of state-subsidized, municipal-owned buses and state-owned and operated rail services within the townships. The existing systems were inefficient and expensive, and Khosa (1995) noted in the 1990s that the majority of township residents spent up to 20 percent of their income on transportation and several hours traveling each day. Because it was initially operated by individually owned eight-seater sedans (e.g. Valiants and Chevrolets), apartheid officials overlooked the industry as a significant threat to other modes of transportation. Market demands facilitated the rapid expansion of the taxi industry as one of the few avenues for capital accumulation for Black owners and drivers, which simultaneously provided more affordable services to commuters (Browning 2001, 2005, 2006). Transportation was also exceptionally political, with fierce racial segregation enforced on both buses and trains. Since the passage of the Motor Carrier Transportation Act in 1930, Blacks had to obtain a special public transportation permit from the Local Road Transportation Board by meeting conditions of employment and residence. The taxi industry thrived as part of the resistance against the apartheid state (Venter 2013).

Several proceedings and legislation facilitated the expansion of the informal taxi industry, leading to deregulation in 1987, which eliminated the restrictions on the distribution of operational licenses. The 1977 Van Breda Commission of Inquiry into the Road Transportation Bill, established to develop mechanisms to prevent bus and train boycotts, found that South Africa had "reached a stage of economic and industrial development which enabled it to move towards a freer competition in transportation" (McCaul 1990: 37). In reaction to this unbridled growth, the 1982 White Paper on National Transport Policy permitted the now infamous 16-seater minibus to enter the market. While the official motivation for deregulation was to strengthen the emerging Black economy, in reality, it also complemented ongoing efforts by the apartheid regime to destabilize South Africa. Deregulation also interestingly mimicked the global trend of market liberalization. As a result, the number of permits issued in 1987 increased from just over 7,000 to nearly 35,000 (Khosa 1991). This influx of operators led to a series of violent conflicts known as "taxi wars", not unlike the "penny wars" in Bogotá, which have marked the industry since (Dugard 2001; for a complete account of the industry in Bogotá, see Ardila 2007). This led to a common assumption that taxis were illegal and operated with impunity. McCaul (1990) provides an alternative account of the taxi industry, disclosing the challenges taxi operators face, as they struggled first against repressive apartheid policy and, more recently against violent mafia-like criminality. Dugard (2001) roots the cause of the taxi violence and the reasons for its continuation in the uncertain socioeconomic conditions of South Africa's democratic transition.

Similar to the experience of informal transportation elsewhere in the world (e.g. dala dalas in Tanzania, colectivos in Argentina, fula fulas in the Democratic Republic of Congo and jeepneys in the Philippines), the minibus taxi industry is popular because of its convenience, frequency and speed as compared to the formal bus and rail alternatives (Schalekamp et al. 2009; Wright 2007b). As a result, excellently engineered South African roads are cluttered with automobile congestion, broken "robots" (traffic lights) and aggressive taxis cutting across four lanes and then stopping suddenly to load passengers before zooming off. In the sprawling landscape of contemporary urban South Africa, the minibus taxi industry has captured the majority of market share against subsidized modes, carrying about 60 percent of trips nationally, because it is considered more convenient in terms of routing and frequency (Clark and Crous 2002). On an average day in Johannesburg, about half of trips are made by private car and half by public transportation, of which three-quarters of Joburgers travel in informal minibus taxis and one-quarter use formal bus and rail services [22]. Cape Town reveals a similar split between public and private transportation but with higher patronage of its rail services [56]. While Cape Town has the highest proportion of household-to-car ownership, reflecting higher average incomes, the alignment between Khayelitsha and the city center is the busiest corridor in the country, carrying 340,000 passengers each day (South African Cities Network 2011). Figure 3.16 indicates the modal split across those South African cities implementing BRT. While there are certainly arguments in support of the minibus taxi industry with proponents describing it as a self-made, Black entrepreneurial venture, for the most part commuters are dissatisfied with the overall quality of service, calling for improvements in the transportation sector (Salazar Ferro et al. 2013).

	Private Vehicles	Public Transport	Minibus Taxi	Rail	Bus	Walking/ Cycling
Cape Town	50	50	30	60	10	3
eThekwini	52	48	30	6	18	-
Johannesburg	53	47	72	14	9	5
Nelson Mandela Bay	47	53	23	5	8	32
Tshwane	66	44	17	7	8	12
Rustenburg	30	70	80	-	10	10

Figure 3.16 Modal split in South African cities.

The high cost of public transportation is a foremost concern, particularly for low-paid workers who spend the largest proportion of their wages on transportation to and from work. While subsidized rail and bus services are usually cheaper than minibus taxi, neither are sufficiently flexible for the dispersed urban landscape. Eighteen percent of South Africans spend 20 percent or more per month on transportation, but this figure increases disproportionately with wages. Fifty percent of those households earning R500 or less per month spend more than 20 percent of their income on public transportation, while 70 percent of households earning more than R6,000 per month spend nothing on public transportation. These costs, compounded with a poor quality of service, contribute towards rising levels of car ownership. Studies reveal that the majority of households earning at least R6,000 per month own a private vehicle. While just 26 percent of households across the country have access to a car, within cities, car ownership remains relatively high with 168 cars per 1,000 persons as compared with just 43 per 1,000 across South Africa[4] (National Department of Transport 2005).

State Intervention in Transportation

The delegation of transportation responsibilities is perhaps more complicated than the origins of the taxi industry: in 1994, public transportation responsibilities were split with national government maintaining responsibility for rail, provincial government managing buses, and cities responsible for municipal planning. The National Department of Transport redefined its role at the strategic level to be an advisory organization, with several quasi-autonomous national government organizations managing the technical decisions (such as SANRAL – South African National Roads Authority Ltd. and PRASA). The rationale behind this transformation assumed that the National Department could be more effective in its devolved state, as "a regulator of bureaucratic detail, a provider of infrastructure, and a transport operator" with a weak focus on policy formulation and strategic planning (National Department of Transport 1996). Broadly speaking, this positioned the national government, and to a lesser extent the provincial government as a "policy maker" and local government became the "policy taker" (van Ryneveld 2006: 16, see also, 2010). National and provincial government rarely appreciated the multitude of needs in the major metropolitan regions and often transportation policies were not put into practice because of a wide range of conflicting statements [54]. This resulted in fragmentation both between national departments and between national and local government [74]. As such, the National Department never directly promoted BRT but rather delegated policy and technical decisions to the cities, only sustaining tangential involvement if cities requested support, pressing for the devolution of the public transportation function to cities, asserting that "cities must choose for themselves" with the national department acting as a "mediator between cities and National Treasury" [38]. These various intra- and inter-city bureaucracies and their role in the circulation and adoption of BRT, will be further unpacked in Chapter 5.

Efforts to regain control of the taxi industry have focused on state regulation and formalization of the taxi industry (Tshwane Department of Transport 2006). State intervention was expected to be comprehensive, covering the administrative, financial and industrial aspects of the industry. One such initiative was the Taxi Recapitalization Program (TRP), first announced in 1999 as a strategy to transform and regulate the minibus taxi industry by replacing old, tattered fleets of 16-seater taxis with a modern, shiny 18- and 35-seater vehicles – purchased by taxi owners after receiving a R50,000 (now R57,000) subsidy for each existing van scrapped. Although the Department of Transport targeted recapitalization of 80 percent of the current taxi fleet, by 2007, only R538 million was spent with another R7.7 billion remaining unallocated. Although the program was for the most part a fruitless exercise – unlicensed drivers still drove recklessly, selectively picking up passengers on impulse and rider satisfaction remained low – importantly for the development of BRT, with the TRP government recognized the significant role of the taxi industry in the transportation network and the national economy, and guaranteed their role in the evolution of the transportation network (South African Cities Network 2011).

Discussions regarding the need for decentralization of transportation functions began in the 1996 White Paper on Transport (National Department of Transport 1996) and the 1999 Moving South Africa Report (Department of Transport 1999), but little progress was made until the passage of the National Land Transport Transition Act (National Department of Transport 2000) (no. 22 of 2000 or NLTTA). The NLTTA provided a set of principles to prioritize public transportation over private car use, and required municipalities to draw up integrated transport plans (ITP) as part of their integrated development plans (IDP). This Act was fairly ineffective since it failed to specifically allocate responsibility for public transportation to cities, and in some cases cities felt that it was an unfunded mandate since there was no direct funding stream. The Public Transport and Action Plan (2007) was instrumental in creating a space for BRT in these conversations. The implementation of BRT offered a chance for cities to advance their involvement in transportation matters by demonstrating proficiency. The passage of the National Land Transport Act (National Department of Transport 2009) (no. 5 of 2009 or NLTA) in 2009 was in response to the introduction of several BRT systems across South Africa, demonstrating the need and opportunity to finally clarify issues of devolution. Cities used BRT to showcase their transportation departments as fully functional.

In March 2007, the South African cabinet approved a public transportation strategy in support of proposed BRT systems (Department of Transport 2007). More importantly perhaps, national government also allocated significant funds for infrastructure and vehicles to support BRT projects. The Public Transport Infrastructure and Systems Grant (PTISG), a conditional grant to municipalities to support the construction of public transportation infrastructure including planning, construction and improvement of new and existing infrastructure and systems, was launched in March of 2005 in partnership with the Department of Transport (National Treasury 2003, 2011). The PTISG is available to the 12 largest South African cities (Buffalo City, Cape Town, Ekurhuleni, eThekwini, Johannesburg, Mangaung, Mbombela, Msunduzi, Nelson Mandela Bay, Polokwane, Rustenburg and Tshwane).[5] The PTISG could be used to finance the construction of public transportation infrastructure, including planning, construction and improvement of new and existing infrastructure and systems. On the other hand, it could not be used to support any operational expenses. The application process typically began with municipal government approving a BRT framework, after which cities applied to the National Treasury for funding. If the framework was consistent with the PTISG guidelines, the National Treasury approved the request for funding [39]. Between 2009 and 2012, expenditure drawn from the PTISG grew from R2.9 billion to R4.8 billion, at an average annual rate of 18 percent, and by 2014/15 expenditure increased to R5.9 billion, at an average annual rate of 6.9 percent (National Treasury 2012). Figure 3.17 indicates the PTISG allocation from 2008 to 2015.

Public Transport Infrastructure and Systems Grant Allocation (in ZAR)	
2008/9	2,919,830
2009/10	2,418,177
2010/11	3,699,462
2011/12	4,803,347
2012/13	4,988,103
2013/14	5,549,981
2014/15	5,870,846

Figure 3.17 Public Transport Infrastructure and Systems Grant allocation.
This figure illustrates the growth of the Public Transport Infrastructure and Systems Grant between 2008/9 and 2014/15. Data drawn from the estimated expenditure in the National Treasury Budget for 2012 (National Treasury 2012).

For many South Africans, the PTISG was the most pressing motivation to build BRT, without which many presume that BRT would not have been possible. While the PTISG may seem like a substantial investment, it is still a tremendous savings over the current costs of subsidizing the bus and rail companies. On average across South Africa, about half of a bus company's revenue comes from national government subsidies with an average of 10 percent increase per year between 2004 and 2010, for a total cost exceeding R2.5 billion per year [92]. Government has tried to institute competition between bus operators and to renegotiate the contracts with bus companies, but without much success. Rail services are in a similarly precarious financial situation, suffering tremendous fluctuation in ridership from about 700 million trips year in the early 1980s to around 400 million in the early 1990s, before rising again to its current level of approximately 600 million. In spite of the R2.5 billion per year subsidy, poor maintenance and under-investment limit the future of the network (South African Cities Network 2011). Transportation experts complain about the financial viability of these subsidized transport systems, with several concluding that it would be more practical to sponsor infill housing or to directly subsidize riders; others saw the introduction of BRT, which promised to reduce the overall costs of operations by introducing a more cost-effective system funded entirely by ticket sales, as the solution.

The introduction of BRT took place concurrently with South Africa's hosting of the 2010 Football World Cup. In May 2004, South Africa was awarded the opportunity to host the 2010 World Cup. This was the first World Cup hosted on African soil and so the event took on an unprecedented level of importance, even as compared with mega-events in other developing nations. The Department of Transport used the excitement and focus of the World Cup to catalyze a "legacy of appropriate, quality transportation services, infrastructure and management expertise" (2006: 35) for all South Africans. The 2010 Transport Action Agenda lists the Klipfontein Corridor in Cape Town as "the first of an interlinking network of BRT lines" (2006: 35). Two other public transportation networks, Johannesburg's Strategic Public Transport Network (SPTN) and eThekwini's North-South corridor focus, are also included as part of South Africa's transportation agenda for the World Cup. Each of these projects served as a precursor to ongoing BRT implementation: Cape Town's Klipfontein Corridor became the MyCiTi route moving along the West Coast completed 11 months after the World Cup; Johannesburg's SPTN became the Rea Vaya BRT network, which was operational for 2010, although it was not yet managed by the official operating company; and eThekwini focused instead on upgrading its rail network, but then in May 2012 approved plans to proceed with BRT. Moosajee rationalizes Johannesburg's decision to build BRT as a means to leave a lasting legacy of improved transportation: "Long after the final whistle has blown, what are the things that the people of Johannesburg will look back and say that these projects may have been done anyway but the pressure of hosting the World Cup really catalyzed some of these legacy projects", she imparted. "And we thought in transport there was a great opportunity to leave a lasting legacy".

This process of devolution is critical in the translation of BRT because it is within this context of municipal centrality that BRT was presented. Cape Town, for instance, used its achievements with MyCiTi to form Transport for Cape Town (TCT) in October 2013 to manage bus, rail and taxi systems. Thami Manyathi, Head of eThekwini Transport Authority, acknowledges that these legislation and frameworks such as the Public Transport and Action Plan, enabled the various cities to innovate and develop their own solutions. Rather than trying to create a universal policy that demands conformity, BRT provided an opportunity for cities to think creatively and innovate [49]. Indeed, BRT is an opportunity for cities to demonstrate their ability to manage the transportation function. By almost all accounts, transportation systems in South Africa were highly problematic – taxis were unregulated and prone to violence, buses were running virtually empty routes and rail was becoming increasingly unreliable. Additionally, buses and rail were draining public coffers through a growing subsidy. In such a scenario, along came BRT, framed as a politically neutral project. Within a few years, cities could transform public transportation and build an entirely new system reaching a stranded populace and demonstrating their ability to manage their function. Many South Africans thought of BRT as a project which "gives life to a lot of abstract ideas and those abstract ideas are not just about transport" but point to larger policy implications. "BRT is a project but it's a project which has a much bigger dimension than even transport – it's about how cities are run" [54].

Negotiating with Taxi Operators

Through BRT, cities can regain control of the informal transit industry by inviting affected operators to become BRT operators. In Cape Town, the city invited existing formal and informal operators to become BRT operators: in Phase 1A, TransPeninsula Investments (TPI), a former taxi association now operates the MyCiTi services along the Waterfront, Kidrogen Vehicle Operating Company (VOC), also previously a taxi association now operates between Dunoon and the Table View, Parklands and Melkbosstrand suburbs, and Golden Arrow bus company was given the contract to operate between the city center and Table View. Similar arrangements exist in Johannesburg – in February 2011, Johannesburg contracted with PioTrans (Pty) to operate Rea Vaya Phase 1A. The company is owned entirely by 313 former taxi operators from 18 affected taxi associations, who are now shareholders in the company and as a result 585 taxis were withdrawn. In April 2014, Litsamaiso was formed from 10 taxi associations (74 percent of the shares) in partnership with Putco (22 percent) and Metrobus (4 percent) to operate services along Rea Vaya Phase 1B. Consequently, 317 taxis were withdrawn. Formalizing the informal operators and building an operating company is certainly the most challenging feature of BRT uptake, and generally

Figure 3.18 BRT in political cartoons.
Political cartoons depict the interpretation of BRT by the taxi industry.

the most unsuccessful component of the various South African projects (see Figure 3.18). It is also a driving force behind the espousal of BRT.

More than a year before BRT first arrived, the Johannesburg Metro began engaging with the taxi industry. In October 2005, the city expanded the area served by its municipally owned bus company, Metrobus, which had historically served the mostly White northern suburbs of Johannesburg into Soweto, the historically Black south-western area. In 2006, the then-newly appointed Member of the Mayoral Committee (MMC) for Transport, Rehana Moosajee began meeting with key stakeholders in the taxi industry (Greater Johannesburg Regional Taxi Council and Top Six Taxi Management as well as the Gauteng Provincial Taxi Council, GauTaCo and South African National Taxi Council, SANTACO) to resolve concerns that the new bus service would threaten their dominance in the area.[6] For the next year and a half, the city continued to engage with the taxi industry who complained that the city was "taking bread out of mouths" and threatening "take your buses out of Soweto or else…" [94]. When Moosajee first heard about BRT in August 2006, she took their proposal to the taxi associations and invited two representatives from each association to join a small delegation visiting Bogotá to learn about BRT [22]. When they returned, the same operators who until that time refused to work with the city to improve public transportation, were interested in understanding how BRT could work in

Johannesburg. Taxi operators remained concerned about the loss of revenue, as their taxi vehicles had to be withdrawn from operations in order to become shareholders in the new BRT operating company. This led to protracted negotiations and many acrimonious meetings (McCaul 2011).

When Rea Vaya Phase 1A began operating on 31 August 2009, no agreement had been reached with the taxi industry. On 28 September 2010, the city signed a 12-year contract with Clidet, an interim company who subcontracted operations to Metrobus to manage operations while negotiations continued with the taxi industry. Negotiations between taxi representatives and city authorities lasted another 14 months, during which affected operators were identified, the bus-operating company was established, the distribution of shares was contracted, reemployment of displaced taxi drivers was arranged, and compensation provided to those no longer employed [5]. On 1 February 2011, the city reached an agreement with the taxi industry and the Clidet shares passed to the new shareholders of PioTrans. Each operator agreed to remove their vehicle(s) from the road and under the government's TRP, and to invest the R54,000 scrapping allowance in the new company as working capital. The bus-operating company is compensated on a fee-per-kilometer basis, and operators are entitled to additional compensation to ensure that their monthly incomes are commensurate with previous earnings. It is important to realize that the city retained all the risk because BRT operators are paid regardless of passenger numbers. The city also assists with purchasing the buses, building the depots and training the new workforce. The affected taxi operators contributed the R54,000 equity, derived from the sale of their minibus in exchange for one share in the operating company [94]. Fanalco, a Colombian-based company with experience managing BRT operations, partnered with PioTrans for three years to help manage bus operations. When their contract ended in February 2014, the technical director of Phase 1A, Javier Cajiao, joined PioTrans to continue managing operations.

The process of engaging operators, which has been more time-consuming than initially anticipated, is illustrative of the sociopolitical aspects of BRT acceptance. Each negotiation process is unique, calling for affected operators to hammer out similar issues, ranging from overcoming the resistance among operators to cooperating with each other and government, determining market shares of the respective models, managing the financial risk versus market share, and delegating supervisory responsibilities equally across the system. As negotiations proceed, they become even more problematic, because each new forum of affected operators wants additional financial rewards. Negotiations along Rea Vaya Phase 1B were more problematic because this route also affected bus operators whose needs had to be accommodated into the business model. Discussions between taxi and bus operators took more than two years and were only completed in April 2014, 7 months after the system began operating. Under that agreement, the city purchased and owned the buses for the first five years, after which they are transferred to the bus-operating company at market value. This agreement more closely resembles that which was decided in Bogotá.

Conclusion

This chapter has explored the global geography of BRT as it flowed from Bogotá to South Africa, to consider why it was adopted and how it was interpreted in South Africa. It first considered the nuances of the structure of BRT and, specifically, Bogotá's Transmilenio before honing in on the specific features of BRT introduced in South African cities. In particular, the process of inviting incumbent informal operators to provide services along the new trunk and feeder system under contract to the BRT authority. As BRT travelled around the world, it was not simply mimicked but rather deployed elsewhere anew, taking on specific attributes reflecting the conditions in the new city.

The theoretical purpose of this chapter was to establish an argument around materiality that starts to explain how and why South African cities adopted BRT. While policy models may seem inflexible as they move through printed documents, the way in which they are described, evaluated and abstracted in relation to needs and conditions of the importing context determines their commensurability. In other words, the strong visual appeal of the material forms in which models travel is not reflected in their application. Rather, as this chapter illustrates, policy models do not move as polished and complete policies ready for duplication in their new surroundings; instead they often mutate and transform in the translation process, becoming something different in their new location. Achieving policy transfer then is not evidence of the superiority of any particular practice or policy, but rather indicative of the local interpretation of the policy model. This explains the variation between South African cities simultaneously implementing BRT – different policymakers made different decisions regarding the policy model. This suggests that policy models are polysemic and rather than replicate; they are modified to suit the needs of each importing jurisdiction.

This chapter has rooted the book in a discussion of materiality. This window into the attributes of policy models recognizes their agentful role in dissemination, but only in reference to the particularities of the people involved (Chapter 4) and the needs and conditions of the importing setting (Chapter 5). The objective of this chapter has been to reveal that BRT is more than just a bus and a busway, but rather a whole host of sociopolitical features which remain open to interpretation in the importing locality. In those cities where engineers managed BRT adoption, the project was more technical; whereas when transportation officials or politicians were managing the process, the project took on a more adaptable nature. Those South African cities that valued the technical, procedural elements of the BRT model over the flexible principles had greater difficulty with the political elements of BRT adoption, while those cities without a strong technical team had problems with various elements of implementation. These arguments speak to the human subjectivity of the policy actors, suggesting that in spite of the ostensible appeal of the Bogotá model of BRT, in fact, it is a locally contingent process furthered by local policy actors that direct its uptake. The next chapter takes on this charge, examining the actors and associations promoting and ratifying BRT, to understand their role in the process of mobility.

Notes

1 There was an earlier attempt to implement a BRT system along Klipfontein Road, which began in 2002, but the plan was not passed in Council and no construction took place.
2 Importantly, the alignment between Soshanguve and the city center is the second busiest rail corridor in South Africa.
3 There are 226 municipalities in South Africa. Since 2000, they have been organized into three categories: Category A including eight metropolitan municipalities – Cape Town, Ekurhuleni, eThekwini, Johannesburg; Category B, also known as local municipalities; and Category C, or district municipalities.
4 For comparison, in 1998, before the implementation of Transmilenio, Bogotá had relatively similar levels of car ownership with approximately 142 vehicles per 1,000 inhabitants (Salazar Ferro et al. 2013).
5 In April 2014, George Municipality was the first city beyond the original 12 to be allocated funds for a BRT system.
6 In 2001, the National Department of Transport formed a representative structure, the South African National Taxi Council (SANTACO), to serve as a unified national representative of the taxi industry. A number of large taxi associations, however, refused to join SANTACO and so its effectiveness has been limited. The Department also established a National Joint Working Group (NJWG) to consult with the taxi industry on legislative and regulatory matters. This group later became the voice for the various BRT projects.

Chapter Four
Actors and Associations Circulating BRT

Introduction

While hundreds of global and local policy actors were involved in BRT circulation in South Africa, certain individuals seem to have played a more dominant, persuasive, dynamic and controversial roll in the process, while others acted as intermediaries better suited to bringing politicians and officials together. The interactions, collaborations and associations among South African and transnational policy actors, speak to wider academic concerns for both the role of individuals as well as networked agency propelling policy exchange. This chapter builds on Chapter 3, which emphasized the role of human subjectivity within materiality in the dissemination of best practice. The cases introduced in this chapter consider the variety of policy actors involved to redress the tendency to assume that policies move rather than are moved. These assertions are rooted in previous analysis of the policy actors who introduce, move and modify extra-local concepts. This chapter opens the black box of policy mobilities by exposing the specific chains of actors and associations spreading BRT through South African cities and, in so doing, furthers my understanding of policy adoption as a feature of the policy actors' engagement with the model, as well as the prospective adopting locality.

This chapter builds an analytic for studying the policy actors and policy networks moving BRT policy across South African cities. The first empirical discussion focuses on the variety of policy actors involved in the circulation of BRT in South Africa. This includes the international advocates engaging with

How Cities Learn: Tracing Bus Rapid Transit in South Africa, First Edition. Astrid Wood.
© 2022 Royal Geographical Society (with the Institute of British Geographers). Published 2022 by John Wiley & Sons Ltd.

BRT circulation, the intermediaries bringing BRT to South African policymakers, and the South Africans building BRT. The analysis of policy actors follows McCann's (2011b) suggestion that policy mobilities involves both supply side (i.e. policy mobilizers) and demand side (i.e. local pioneers) engagements, but adds a third dimension: the intermediaries who translate and localize the concept between global and local actors. In concluding this section, however, it becomes evident that while individual policy actors can mobilize and implement innovation, a diversity of actors advance policy mobilities.

This leads to the second empirical section of the chapter, which builds on the detailed exploration of various policy actors to explore the global and local associations and interactions between them, to understand where agency lies in policy mobilities. This section explores the involvement of formal associations (i.e. Institute for Transportation and Development Policy [ITDP]) and informal networks (i.e. friendships formed on study tours) in BRT adoption. I consider these actors within their networks to address the assumption that policy mobilities is a straightforward bilateral exchange between one locality and another, or even between international advocates and local implementers. My consideration for the power dynamics of networks exposes policy mobilities as being both individually- and institutionally-driven and equally, internationally pushed and drawn-in; and adds an important understanding of the flexibility and durability of relational networks in shaping policy mobilities. This chapter therefore offers an opportunity to understand more precisely who moves policy and how they do it, within a larger conversation regarding why cities employ mobile knowledge differently. That is, I explore the relational attributes through which different actors and associations generate divergent understandings and procedures, thereby leading to differences in uptake. Such claims speak to the wider aims of this book to explore the significance of power relations and personality in advancing universal ideas.

An Analytic for Studying Policy Actors

A policy actor can be anybody involved in the consideration, discussion and analysis of policy, and is not necessarily limited to stakeholders or elected officials. Much of the focus in the literature has been on the "policy mobilizers" (Wood 2014b), or the policy and urban planning professionals, practitioners, activists and consultants working as "policy entrepreneurs" (Dolowitz and Marsh 1996), mobilizing knowledge across geographic, historical and institutional contexts. "Transfer agents" (Stone 1999, 2004), "intermediaries" and "cosmopolitans" (Sutcliffe 1981: 173) are the actors shuttling policies and concepts between places. These actors acquire, test, convert, store, and finally disseminate socially produced, circulated forms of knowledge shaping various networks, policy communities and institutional contexts. In the arguments that follow, I reveal that policy mobilizers may not usually be affiliated with either the exporting or importing jurisdiction, and in many instances, they are "policy gurus" (Peck 2011b) with a variety of expertise. Philosophers, political economists, philanthropists (Rose and Miller 1992) and "chance contacts" (Mossberger and Wolman 2003) also act as policy mobilizers, when they are instrumental in planting ideas that lie dormant until local policy actors engage in their prospective evaluation. Historically, policy mobilizers were the international planners working in the colonial service spreading British models around the globe (Cherry 1981), the developers experimenting with a new town planning scheme commonly used in another locality (McCann 2011b; Shamsudin 2005), and the civil society workers disseminating best practice through transnational policy networks (McFarlane 2009).

Employing outsiders can help generate a fresh perspective, especially on highly politicized issues. Hence, the rapid proliferation of the "global consultocracy" of private policy consultants propelling policy migration (McCann 2011b). In McCann's (2011b) study of consultants, he emphasizes the role of "incoming policy consultants" who bring knowledge from one city to another, and "outgoing policy consultants" whose practice involves disseminating the successes of one city to people elsewhere. It is important to realize that these actors are not simply vehicles for impartial, rational knowledge as the policy transfer literature might assume, but rather their actions are always inherently political, and perhaps at times self-serving. Policy mobilizers often try to "sell" rather than "tell" stories of policy innovation (Peck 2011b; Wolman 1992), and they can easily introduce particular policies and bring certain cities into conversation with each other while pushing others apart. Figure 4.1 provides an illustration of the types of policy actors presented and analyzed in the literature.

This chapter will explore how some actors are instrumental in planting ideas that might lie dormant, while others engage in their prospective evaluation and application. Individual policy actors cannot simultaneously create, impart, mobilize and approve transnational models. Rather, it is through the messy and tangled webs of relationships linking different kinds of actors through the distribution of international innovation, that stretch urban policy across the globe. In moving innovation across jurisdictional boundaries, policy actors form associations and networks among themselves, with policy actors in both exporting and importing localities, and even with the mobile policy. These networks of policy actors legitimize blanket policy by giving it both local and transnational salience. They form, at a particular juncture, to disperse innovation and then often fall out of contact once the transmission is completed.

Types of Policy Actors

"Intermediaries" and "cosmopolitans"	Sutcliffe (1981)
"Idea brokers"	Smith (1991)
"Philosophers, political economists, philanthropists"	Rose and Miller (1992)
"Policy entrepreneurs"	Dolowitz and Marsh (1996)
"Dolowitz and Marsh model": "elected officials, political parties, bureaucrats/civil servants, pressure groups, policy entrepreneurs and experts, transnational corporations, think tanks, supra-national governmental and nongovernmental institutions and consultants"	Dolowitz and Marsh (1996, 2000)
"Transfer agents": "non-state actors involved in think-tanks, research institutes, consultant firms, philanthropic foundations, university centres, scientific associations, professional societies, training institutes, NGOs and pressure groups"	Stone (1996, 1999, 2002, 2004, 2010)
"Global Intelligence Corps"	Olds (1997, 2001)
"Chance contacts"	Mossberger and Wolman (2003)
"Consultancy firms, transnational institutions, policy networks, think tanks, governmental agencies and professional institutions"	Peck (2003)
"Mediators"	Osborne (2004)
"Global consultocracy" including "incoming policy consultants" and "outgoing policy consultants"	McCann (2011b)
"Policy gurus"	Peck (2011b)
"The redevelopment professional" – "part economist, part engineer, part planner, part marketing executive"	Ward (2011)
"Politicians, policy professionals, practitioners, activists and consultants"	McCann and Ward (2011b)

Figure 4.1 Types of policy actors.
This figure lists the various state and non-state actors assembling, mobilizing and adopting global knowledge.

The next section analyzes these various actors forming, informing and sanctioning BRT in South Africa. It traces the introduction and approval of BRT projects and policies through the local and international policy actors, to recover the role of individuals that move policy. This exploration of the different kinds of policy actors managing the migration of BRT concepts expands our understanding of the varied direction, speed and influence of global and local influences shaping the South African city.

Redefining the Role of Policy Actors

In August 2006, two international consultants, Lloyd Wright and Todd Litman, visited Cape Town, eThekwini, Johannesburg and Tshwane spreading information about BRT to city officials and politicians (Litman 2006). Lloyd Wright is an international transportation expert advocating for BRT implementation across the globe and a consultant affiliated with the ITDP. Todd Litman is a researcher based in Vancouver, disseminating innovative transportation solutions around the world, who happened to be in South Africa for the Southern African Transport Conference and joined Wright to promote BRT. Their visits with city officials and BRT presentations were noted as a fundamental moment in the transfer of BRT across South Africa. The impact of these meetings was demonstrated when three years later, in August 2009, Johannesburg commenced operations on its Rea Vaya and, in May 2011, Cape Town launched MyCiTi.

While it appears like an ordinary case of best practice introduced by international policy consultants, something that has already been explored elsewhere (Dolowitz and Marsh 2000; McCann 2011a, 2011b; Stone 2004), upon further consideration, the introduction of BRT exposes locally contingent processes which have yet to be fully theorized. In my interviews, there were strong arguments in support of the need for charismatic policy mobilizers [41; 47; 95], while others disputed such claims, reasoning that synergies between local policy actors were responsible for BRT adoption [17; 55; 68]. There were also debates regarding the value of international voices in the dissemination of BRT [26; 27], as well as the converse praising the astuteness of South African experts [22; 55]. South Africans sometimes called the international consultants, "missionaries getting their story out to the world" [60], while others deduced that BRT will never succeed if it is internationally driven [27]. Many South Africans attributed BRT in South Africa to the efforts of particular policy mobilizers like Lloyd Wright [8; 11; 17; 42; 47; 54; 60; 61; 64; 68; 70; 77; 89]. The evidence suggests a need to understand the different kinds of policy actors. Perhaps certain policy mobilizers introduce policy models, some actors serve as intermediaries bringing it into conversation with the local implementers, and still others localize and adopt it. In making the case for studying the typologies of policy mobilizers, I consider the actors and associations moving BRT through South African cities, in particular contemplating the role of policy mobilizers, intermediaries and local pioneers in creating, communicating, circulating and embracing BRT. Figure 4.2 illustrates various roles and tasks for each type of policy actor and names particular individuals involved in these tasks within the South African case study. These categories emerged directly from comments made by interview respondents.

	Disseminate successes (e.g. Bogotá)	Translate into international model (e.g. BRT model)	Generate awareness in importing jurisdiction (e.g. study tours and legislation)	Translate awareness into action (e.g. from SPTN to BRT)	Implement project (e.g. build Rea Vaya and MyCiTi)
Policy mobilizers (e.g. Enrique Penalosa; Lloyd Wright)	X	X			
Intermediaries (e.g. Philip van Ryneveld)			X		
Local pioneers (e.g. Rehana Moosajee; Bob Stanway)				X	X

Figure 4.2 BRT policy actors and actions.
These categories emerged through interviews with South Africans in which I asked them about the different types of policy actors involved in the circulation of BRT concepts.

The remainder of this section utilizes these categorizations – policy mobilizers, intermediaries and local pioneers – as an analytic through which to explore the types of policy actors and the connections between them mobilizing BRT circulation.

Policy Mobilizers of BRT Circulation

A host of local and international advocates plant ideas like BRT in the minds of local pioneers. In the case of BRT circulation in South Africa, policy mobilizers like Enrique Penalosa, the mayor of Bogotá responsible for implementing Transmilenio, was described as "a messiah" with "all the answers and success stories of Bogotá" [87] who "can sell anything" [53]; while Lloyd Wright was often recognized for bringing BRT to South Africa. Penalosa's presentations focused on improving sidewalks to encourage walking and building new bikeways to inspire cycling, as well as BRT. He explained that his motivations were to build a more democratic and sustainable city, not just spread his success stories [72]. "What a powerful idea" suggests Paul Steely White, Executive Director for Transportation Alternatives and formerly the Africa Regional Director for ITDP, "instead of transportation infrastructure being used to fragment communities, it can be used to fuse communities" [41]. The actors "bring a Southern success story that resonates with African decisions-makers who are facing tough choices about the future of their cities" (White 2003).

South Africans describe Enrique Penalosa as "dynamic" [49], a "showman" [38] and a "messiah for livable cities" [17], who "brings rain to the desert" [59]. International consultants who share this disposition describe Penalosa as an "energizer bunny" with "pizazz and panache" that builds "Bogotá up as some miracle" [41]. Penalosa is praised by South Africans for being an essential element of BRT exchanges. He is recognized for connecting South Africans with a successful example and convincing skeptics that BRT was possible. Ibrahim Seedat, Director of Public Transport Strategy at the National Department of Transport, an intermediary explains, "You need somebody like Ken Livingstone in London or Enrique Penalosa in Bogotá" because "there are so many challenges that you get swamped" [38]. Local South African politicians, like Rehana Moosajee, Member of Mayoral Committee (MMC) for Transport in Johannesburg from 2006 to 2013, acknowledge the importance of having these interactions with policy mobilizers like Penalosa "who had been through this and understood what it was going to take" as essential element for her learning [22].

Another important figure in the diffusion of BRT through South Africa and perhaps the most polarizing figure in the story of BRT is Lloyd Wright, an international consultant with ITDP, founder of VivaCities and author of the BRT Planning Guide (Wright 2007a). Wright is the individual often credited with selling BRT to South Africans. He is a classic example of a policy mobilizer, a smart and convincing consultant with expertise on the universal model as well as familiarity with the local conditions, and habitually the-right-man-at-the-right-time, able to link international and local policy actors, thereby furthering the program [11; 17; 70; 89]. Madeleine Costanza, who worked for the International Institute for Energy Conservation (IIEC) from 1998 to 2002 and was instrumental in the preliminary broadcast of BRT to South Africa, describes Lloyd as "the real pioneer of BRT in South Africa because while a lot of bus and taxi improvements were envisioned, he was really the one that made the experiences and costing and finer details of BRT known to South African transportation professionals, including me" [23]. Many interview respondents shared similar experiences of learning of BRT from this policy mobilizer, and I therefore conclude that Wright played one of the biggest roles in introducing South Africans to BRT, by bringing both the details of the concept as well as a host of international examples to South African policy actors.

These policy mobilizers are persuasive, able to convince intermediaries to influence local policy actors to undertake major capital and political exploits. Paul Browning, a consultant working with minibus taxi operators in Nelson Mandela Bay, suggests that part of Lloyd's charm comes from his honesty. "Lloyd looks you straight in the eye and says you can trust me and he is the expert so you do"

[95]. Another respondent uses similar language to illustrate Penalosa's panache. "Enrique looks them straight in the eye and says this is one of the best thing you could ever do for your political career and then while it's politically tough to implement, ultimately the political benefit is worth the investment" [41]. These international consultants are politically astute and topically knowledgeable, but they use their charm to forward best practice. One Capetonian engineer recalls, "Enrique Penalosa doesn't need the materials, he is Enrique Penalosa. He has the passion and he's done it. But somebody like Lloyd Wright – he has the expertise but I imagine that he said 'look at all this material evidence'" [59]. Many South African praised these policy mobilizers rationalizing that their involvement "made all the difference" [22] without which it would have been difficult to sustain BRT implementation [47].

Not all intermediaries play such an obvious role in policy mobilities. Whereas Lloyd Wright assisted South Africans with the political maneuverings of BRT adoption, technical consultants were brought in to share their knowledge of construction, financing, operations and management. One international transportation engineer, Pilo Williumsen, hired in 2003 to do a feasibility study for the proposed Klipfontein Corridor in Cape Town, distinctly remembered his first order of business was to unearth knowledge of the minibus taxi industry – who drives the taxi, who rides in the taxi and how the services operates (Willumsen and Lillo 2003). He was astonished to find that none of his local colleagues had personal experiences riding in taxis and not one had a personal relationship with the minibus taxi providers. As an outsider, he was void of the prejudices and stereotypes that many of his South African collaborators held towards the industry. Thus, early one morning, he arose to experience the morning commute between the south-western townships and the city center. Many of his experiences fit the preconceived typecasts – the taxi was overloaded with passengers rushing to work and broke down along the highway, halfway to town – but what he found fascinating was the speed in which another taxi arrived to fetch the passengers and complete the journey. While some locals may have seen a malfunctioning vehicle, he saw a highly functioning system, one that had to be cultivated through the implementation of BRT. He utilized his time on the road to learn about the taxi industry and build a rapport with taxi drivers and industry officials [51]. This stranger was able to build a relationship with the taxi industry in a way that was not possible for locals. His external outlook also added unique knowledge of both the global and local context, making his services indispensable.

Policy mobilizers are usually paid for their knowledge delivery services by either the prospective importing locality or international agency. Through both formal invitations as well as personal connections and friendships, policy mobilizers fly into a new city, introduce innovation and then step back and let the local city develop the project, before moving on to the next project in another city. Policy mobilizers rely on their networks as well their reputation and the success of the policy model, to ascertain new opportunities to disseminate best practice.

The process in which policy mobilizers parachute-in and parachute-out and try to change the world in two weeks, leaves insufficient time to acquire adequate knowledge of the local complexities, and thus they do not understand how to tailor best practice in one place to suit an alternative context. Wright acknowledges one of the key things he learned from his involvement in South Africa is the importance of embedding oneself in the local context rather than making supplementary visits, which are both time consuming and cumbersome and still often results in problematic implementation [27]. To ensure a more successful outcome, between 2007 and 2009, Wright broke from his usual role as a policy mobilizer and embedded himself full-time leading Cape Town's BRT project, MyCiTi. However, because "Lloyd's strengths are not in implementation but in getting the idea sold and doing some initial conceptualization" [54], his employment in Cape Town led to a host of political complications, including a massive underestimation of the total cost of the project.

The involvement of international consultants in BRT projects was heavily scrutinized with one interview respondent maintaining, "the missionary is always selling the idea and not the critique and is often selling some problematic ideas as well" [60]. Policy mobilizers often only share the successful translation, not the difficulties transferring a model across localities. Whereas they may hold considerable knowledge of universal examples, they rarely impart the complete picture of policy experiences. Those skeptical of the international involvement in South African BRT realization suggest that the policy mobilizer's ability to broadcast concepts like BRT comes from their personality. One consultant in Cape Town recalls how the policy mobilizers "jumped in with very little sensitivity and consequently was able to have an impact but there were more than enough knives around when the right opportunity came" [60]. Here he is referring to the rising cost of Cape Town's MyCiTi system, which in 2009 resulted in the termination of Lloyd's contract as well as the resignation of Eddie Chinnappen, Cape Town's Director of Transport. Experiences such as these suggest that the role of international consultants in local policymaking is more complicated than a simple fly-in, fly-out practice because policies, and the costs of implementing them, are shaped by the local socio-political context. A policy mobilizer rarely sticks around long enough to calculate the costs of a BRT system, much less, to see the budget inflate. As such, intermediaries are useful in bridging the gaps between international and local policymakers.

Intermediaries of BRT Circulation

In my interview with Andrew Boraine, City Manager of Cape Town from 1997 to 2001 and CEO of Cape Town Partnerships from 2003 to 2013, I asked, "Did Lloyd Wright introduce BRT to Cape Town?" Boraine responded by suggesting I speak with Philip van Ryneveld because "he was instrumental in bringing Lloyd out and working with him in the early days of conversations" [17]. Intermediaries like van Ryneveld are necessary to navigate the translation of BRT from an international concept to a municipal project. Policy flows are often riddled with human and material obstacles – perhaps a politician is unwilling to risk his position on an untested investment or an engineer cannot maneuver the technical adjustments of BRT platforms along a narrow street in an historic district – and yet this tends to be ignored in the literature. This section exposes the role of intermediaries like Philip van Ryneveld in translating ideas for the local context, by linking international policy advocates with those South Africans able to implement BRT.

Philip van Ryneveld was the Chief Financial Officer (CFO) in Cape Town from 1997 to 2001 and since then runs a consultancy specializing in public finance management and economic development. Although not currently a politician or official with the city, van Ryneveld played an integral role in introducing BRT to Cape Town, spreading the concept through his dynamic personality and expansive political connections. Van Ryneveld is commended for introducing Lloyd Wright to then-Mayor of Cape Town, Helen Zille who subsequently initiated the MyCiTi project. Van Ryneveld recounts his role in the story of BRT in Cape Town beginning in July 2006 in Johannesburg:

> I was talking to [a colleague at the South African Cities Network] one day in July 2006 about my growing interest in transport as it relates to the devolution of built environment functions and she said to me, "There is this guy Lloyd Wright in town and maybe you should go and meet with him?"
>
> I said, "Who is he? Why would I meet with him?"
>
> And she said, "Go and meet with him and see what he has to say".
>
> So off I went to Melville that evening to meet Lloyd and he explained the BRT concept to me...

Here, van Ryneveld draws on his informal network which connects him with the South African Cities Network (SACN), a learning network linking the nine largest South African metropolises. Van Ryneveld also relied on his personal connections with Jeremy Cronin, Deputy Minister of Transport from 2009 to 2012 and Chairperson of the Parliamentary Portfolio Committee on Transport from 2003 to 2009, another critical intermediary, who first introduced van Ryneveld to the BRT concept at a meeting regarding the devolution of built environment functions and transportation. Cronin mentioned BRT and gave van Ryneveld a CD, which Cronin had received from Walter Hook, CEO of ITDP from 1993 to 2014, with one of Wright's BRT presentations – although van Ryneveld never followed up until he met Wright later that year.

Knowledge of BRT was only activated when van Ryneveld met Wright in Johannesburg. The role of van Ryneveld was to mediate between the international consultants and South African policy actors. When at the end of the meeting, Wright said, "I'd like to take this to the MEC in the Province and we can try and get this further", van Ryneveld responded, "Don't take it to the province, there is no way that you'll get public transport done through the provinces, the only organization that can achieve something like this is the city government. You have to take it to the Mayor". Van Ryneveld again relied on his personal network; he knew Helen Zille from their time working for Black Sash (an anti-apartheid movement) in the 1980s, so van Ryneveld called the Mayor and said:

> "Helen, there is this concept of BRT and I think you should hear about it" and she said, "Let's have a meeting".
>
> [Philip] said, "Well, I don't know much about it but the next time this international guy Lloyd Wright is in town, we should meet then".
>
> She said, "Yes, let's do that. Next time Lloyd is in town, get in touch with me and we can have a meeting" [54].

On 23 January 2007, Lloyd Wright and Ibrahim Seedat, Director of Public Transport Strategy at the National Department of Transport, came to Cape Town to meet with Mayor Helen Zille. Wright presented the story of BRT and Bogotá's successful launch of Transmilenio, and the Mayor responded with interest in undertaking BRT in Cape Town. Van Ryneveld's central role in organizing an interaction between those with the idea and those with the power to implement the innovation is critical to policy mobilities.

Intermediaries like van Ryneveld play an important role in advancing the execution of disseminated notions. They often negotiate more subtle interventions, arbitrate handshakes between international and local policy actors and then step back to allow those with power to make decisions. Intermediaries are often consultants whose paths and activities overlap with both international policy mobilizers and resident politicians and practitioners. They might play multiple roles in the transfer process, providing introductions between policy mobilizers and local pioneers, acting as observer and later critic of the implementation process, and even drafting government documents and legislation regarding the intervention. Their reward comes later, perhaps in the form of a future tender. Intermediaries generally lack formal political or official capacity but rather link the international with the local. Lloyd Wright remembers van Ryneveld's involvement as critical in implementing BRT in Cape Town [27]. He used his personal network from his days in the anti-apartheid movement to contact Helen Zille, who would ultimately be able to move the project along and, including Ibrahim Seedat from the National Department of Transport, was crucial in building synergies between the national and local levels of government [38]. Alone, van Ryneveld could neither sell an idea like a policy mobilizer nor fund one like a local pioneer, but through his active engagement with both international and local policy actors, he played a critical role in policy adoption. However, even the most talented intermediary still lacks the ability to make policy.

The involvement of intermediaries is a particular South African condition in which antiapartheid activists were instrumental in forming the new post-apartheid government, and when their terms concluded remained involved as consultants. Moreover, because so many new ideas and voices were urgently needed, those with knowledge and experience moved between cities and across spheres of government collaborating to build a better future. Pauline Froschauer, another example of a transportation planner who has been involved in multiple aspects of BRT, embodies the term intermediary. Froschauer and her firm, Namela, designed, developed and then managed the construction of the Rustenburg BRT system. Like many intermediaries, she fills multiple roles at different stages of the learning process. Previously, Froschauer was involved in the development of the National Department of Transport Public Transport Strategy and Action Plan (2007), as well as the initial conceptualization of Tshwane's BRT system. More recently, she managed a consortium of consultants including Walter Hook and Lloyd Wright to provide support to the National Department of Transport. Froschauer and her consortium are responsible for reviewing South African cities' application of BRT principles. Froschauer, however, must be careful not to get embroiled in conflicts of interest – as overseer of the National Department of Transport consortium, for instance, she excludes herself from the assessment of the system in Rustenburg [75]. In addition to participating in implementation and national policy development, her role connecting international and local consultants as well as other intermediaries, is emblematic of the variety of roles one policy actor can play in the circulation process.

Local Pioneers of BRT Circulation

Local pioneers are the actors who localize and apply the innovation. Their connection through intermediaries to international policy advocates distinguishes them from ordinary policymakers. Interview respondents rationalized that politicians and local pioneers are "absolutely critical" when driving the application of mobile knowledge [62]. South African policy actors across the cities indicated that having local political and technical support for BRT, was fundamental to drive the project from concept through to completion [50; 60; 70].

Politicians are one type of local pioneer. International consultants and intermediaries must "sell [BRT] to politicians" because policy adoption is "only achieved alongside political will" [45]. In return, politicians might take "a leap of faith" in pushing ahead with the BRT project [17]. Jeremy Cronin confirmed that local political champions were essential in shouldering the responsibility for BRT in South Africa [62].

One particularly impressive politician is Rehana Moosajee, a figure praised across interviews – even by her political opponents and the taxi industry – for her pioneering devotion to BRT realization [8; 31]. She was after all the first politician to stand up to the politically powerful taxi industry, even as President Zuma's election campaign promised to cease BRT construction. Moosajee remembers that she first learned of BRT in 2006 at a presentation by Lloyd Wright and his colleague, Todd Litman, who travelled around South African cities spreading the BRT gospel. Moosajee describes the meeting as serendipitous because she was not meant to attend the presentation they made in Johannesburg.

> I stumbled upon it because I was between meetings and I wanted to get something to eat and they were there. So I went there to get something to eat and my team was actually making the presentation and then somebody quite cleverly said, well while the MMC is here should we just talk about this thing and so literally there was an element of serendipity. At a personal level, I am quite inclined to believe that everything happens for a reason and there are many things in this project that suggest that the timing was right and the necessary players were brought together at the right time to make this happen [22].

This meeting was particularly unusual because it brought the political and technical leaders of municipal transportation together, and provided a platform for them to discuss the possibility of welcoming BRT.

Andrew Boraine suggests policy adoption requires bold, forward thinking politicians such as Helen Zille, Mayor of Cape Town from 2006 to 2009 and Premier of the Western Cape from 2009 to 2019. Boraine continues, "I don't think she really understood much about those systems and it wasn't high on her agenda but she got it quite quickly. [Lloyd] was able to sell it to her and that helped. If you get a mayor behind you, it always goes somewhere or it has more of a potential" [17]. Lloyd Wright confirms Boraine's experiences, of the value of getting the support of local politicians. He remembers his assessment of local South African politicians from his involvement in BRT advocacy.

> We live in a global world but it is still surprising that decision makers have never heard of BRT. Rehana Moosajee when we first met her had never heard of that concept but she was immediately taken by it. The same thing with Helen Zille who was then the Mayor of Cape Town and now the Premier of Western Cape. She actually stopped us about 10 minutes into the presentation and said "Ok, I'm sold. How do we do this?" And that's for a lot of these sustainable transport concepts, not just BRT but also pedestrian and bicycle improvements. They make a lot of intrinsic sense to people who are generalists. These politicians are not transport specialists [27].

Busy politicians like Moosajee and Zille do not read policy documents and deliberate on how best to implement those ideas; they rely on succinct, sellable ideas. Personal relationships are therefore particularly important for such politicians with little time to study the details of a policy or program. They rely on their network of trusted staff as well, as personal relationships with policy mobilizers, to interpret and frame best practice suggestions. Politicians are thus collaborators managing complex webs of relationships.

Politicians also work in concert with other politicians from other levels of government, as well as with intermediaries. Jeremy Cronin, for example has been quite successful at choreographing BRT support from the top level, both in terms of negotiating the devolution of public transportation responsibility to cities as well as in maneuvering the creation of the Public Transport Infrastructure and Systems Grant (PTISG) at National Treasury (see the 'State Intervention in Transport' section). In addition to his role as Deputy Minister of Transport, Cronin also serves as an intermediary, inviting city officials to meet with Bogotá city officials [60; 62]. Cronin was instrumental in bringing Walter Hook, CEO of ITDP, to South Africa on many occasions and managing his communication with local politicians. Walter Hook, in return, commends Cronin as key to his awareness of contextual issues such as the transformation of the taxi industry [73]. Cronin is often credited as the politician with the most sustained concern for public transportation and, as was previously detailed, he was instrumental in introducing van Ryneveld to BRT and Lloyd Wright's involvement in Cape Town's BRT project could also be traced back to Cronin [54].

Local implementers are another category responsible for localizing the innovation and transforming tacit knowledge into explicit knowledge. They are also pioneers because their role mandates them to change their ongoing practice and experiment with an imported idea, which at times brings risks to their position. In Johannesburg, this meant Bob Stanway had to stop construction on the Strategic Public Transport Network (SPTN) and start construction on the Rea Vaya network [8]; in Cape Town, this meant Ron Haiden had to turn his sights from building public transportation interchanges towards building the MyCiTi stations [55]. Whereas policy mobilizers actively engage in publicizing the ideas, local implementers are the actors responsible for bringing the idea to fruition.

Bob Stanway was project manager for Rea Vaya Phase 1A from 2006 to 2009 and a proponent of public transportation in Johannesburg for several decades. He has been an influential and intellectual force in public transportation, employing various concepts learned from his London-based education and work experience to Johannesburg. Through his consulting firm, Stanway Edwards, Stanway pursued an agenda of public transportation, urging the construction of a light rail network in Johannesburg, when most transportation engineers were firmly supporting highway construction and automobile dependency [8]. In this capacity, Stanway also influenced several South African transportation specialists including Colleen McCaul with whom he worked with on Rea Vaya [5]. Stanway is described by colleagues as "key to transport in Johannesburg" [44], and "key to the fact that today we have the Rea Vaya system running" [36], with "vital technical expertise for the project" [22]. As local pioneer, however, Stanway was tasked with mediating between the implementers of the project, in this case the Johannesburg Development Agency (JDA), the city's development agency, the city's transportation department, and the international consultants. Stanway had "the burden of having to make the decision to go for BRT" and used his personal network to "negotiate with local stakeholders" [53].

Ron Haiden, the Manager for Infrastructure and Development in Cape Town, is another remarkable local pioneer. Several interview respondents give Haiden the credit for implementing the MyCiTi system, including Eddie Chinnappen, former Executive Director of Transport, who attributes the route up the West Coast along the R27 to Haiden's knowledge of the availability of land, and Maddie Mazaza, Cape Town's Executive Director of Transport, recognizes his ability to navigate politics [70; 89]. "You need a special person like Ron who will get involved in the art and design", asserts John Jones, Director of Engineering at HHO Africa. "He is also a detailed guy who can be in both places. Without that sort of detail, a lot of things might not have been built right" [87]. When I asked Haiden how he learned of BRT, he exclaimed, "I just knew because of almost thirty years planning transportation projects in Cape Town. That kind of knowledge is fundamental on a project as complicated as MyCiTi" [55].

Industry officials are also integral to policy change. Indeed, the support of minibus taxi operators was fundamental to the successful introduction of BRT. Eric Motshwane's transition from Chairman of the Greater Johannesburg Regional Taxi Association, an industry opposed to BRT, to Corporate Director of PioTrans, the operating company for Johannesburg's Rea Vaya Phase 1A, is of critical importance to the fulfilment of the project. Motshwane is a local pioneer, who bravely took on the taxi industry, motivating the project along in spite of the threat of violence. He told me that his fear of losing credibility with the taxi industry was surpassed only by his belief that BRT was the right choice for the future of the industry [94]. Motshwane's boldness is praised in a number of interviews for being "sensible" and "patient" [95] and most notably by Moosajee, who commends the "visionary leadership" of the taxi industry who risked their lives "in the interest of the industry", explaining that "issues of collaboration ... made all the difference" [22].

Local pioneers are also the ones who often pay the highest price when mobile policies are insufficiently localized or inappropriate altogether. Eddie Chinnappen was Executive Director of Transport in the City of Cape Town between 2006 and 2009, and as a trained transportation engineer worked in every sector of public transportation in Cape Town for 23 years. In 2009, Chinnappen resigned as MyCiTi project costs escalated from R1.4 billion to R4.1 billion. Local pioneers are often the ones challenged with localizing the project – in this instance, determining the financial cost to the city of building BRT – and hence often blamed for any errors along the way. Chinnappen affirms, "I am not sad. I don't hold grudges against anybody. ... As executive director at that time, I took responsibility for any misconceived ideas of the project" [89]. In spite of his termination, Chinnappen was applauded across a number of interviews for his dedication to transportation improvements in Cape Town [68; 86; 88].

This section has detailed the types of policy actors involved in the circulation of BRT in South Africa, drawing attention to a more diverse range of types of individuals than the literature has so far acknowledged. The next section focuses on the formal networks on which various policy actors rely and the new associations that emerge throughout the learning process.

Learning through Networks

Policy actors work through a variety of professional and nonprofessional, local and international, governmental and nongovernmental organizations, sometimes referred to in the literature as policy networks. Generally, a network is formed by "a series of linkages – formal and informal, permanent or ephemeral – which bind together entities that are geographically distant, either in a single country or across boundaries" (Saunier 2002: 512). These networks act as representatives, proxies, vehicles, drivers, and negotiators, mediating and spreading knowledge around the world through conferences, study visits, consultancy and reports. Sometimes ordinary policy actors become global policy disseminators by using these organizational platforms to distribute knowledge.

These relationships can be difficult to uncover and complicated to unravel, so I asked South African interview respondents, "Who told you about BRT?" to understand the range of actors that shaped BRT adoption. Figure 4.3 illustrates the importance of local practitioners and colleagues rather than international experts in the reception of BRT – although these global voices can be more vociferous than those of South African actors. Classic social network analysis represents the patterns of city-to-city visits as nodes and ties to symbolize the individual policy actors and the relationships formed (Campbell 2012; Said and Herzog 2011), and Figure 4.3 enables a consideration of the variety of players and the environment through which BRT information moved. It indicates that the strength of the South African policy network is in local practitioners, who are both topically knowledgeable and familiar with current conditions, to utilize the information from policy mobilizers to build BRT locally. They do so through various international and South African associations and networks.

Edgar Pieterse, Director of the African Centre for Cities at the University of Cape Town in his explorations as both a city official and academic, reasons that "If the learning is solely individually-driven then it won't get very far because the sticky point of implementing any policy is the challenge of putting together action networks. ... the key thing is to find people who can facilitate that

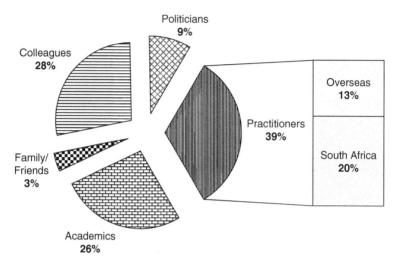

Figure 4.3 'Who told you about BRT?'.
Interview responses to the question, "Who told you about BRT?", reveal that nearly 40 percent of South Africans learned about BRT from practitioners, of whom two-thirds are based in South Africa and only one one-third come from overseas. This indicates a greater proportion of learning took place across South Africans than between South Africans and international experts. Nearly 30 percent indicated that they learn alongside colleagues through ordinary work experiences. These factors suggest a diminished role for international influences and rather indicate a need to understand the ebbs and flows of knowledge taking place within South African cities.

individual and/or group-based learning..." [66] Knowledge based solely within structured networks, however, can be difficult to access and as Bea Drost, a policy analyst at the National Planning Commission, reveals, it becomes "a task of knowing people rather than information" [35]. These observations illustrate a "culture of learning" [46] in which politicians and officials collaborate. The following section examines this culture to understand the role of formal networks like ITDP and informal associations formed while overseas on study tours, in ongoing BRT adoption.

Networks of Internationals

Sustainable transportation advocates like ITDP, an NGO based in New York, Embarq, the WRI Center for Sustainable Cities in Washington, DC, and the Volvo Research and Educational Foundation (VREF) in Gothenburg, Sweden, are the principal sponsors of BRT, moving knowledge and experience between Bogotá and South Africa. These associations often provide training manuals, funding for study tours and the names of architects, engineers and planners with topical expertise and skills. These examples of "intergovernmental diplomacy" (Saunier 2002: 509) and "transboundary connections" (Saunier 2002: 510) manage and sustain collaborations by promoting innovation, strategic thinking and friendship between cities and among various spheres of government. They provide a forum for both formal (i.e. seminars and reference groups) and informal (i.e. networking) engagements, and are not usually formed for the explicit purpose of delivering a particular innovation like BRT, but rather rely on pre-existing relationships across the network to inspire local support for any given policy or project.

ITDP was founded in 1985 by a group of sustainable transportation advocates to replace the export of the American model of car dependency with bicycle schemes (ITDP 2013). Two decades later, ITDP has morphed into a "BRT propagation machine", helping cities with all elements of BRT from operations to infrastructure [41]. The NGO has been credited with "prying open the doors" [23] and being "seminal in changing the minds of South African transport professionals and political managers", because of their extensive efforts to publicize Transmilenio through professional presentations and study tours orchestrated by Penalosa [30]. The organization became involved with BRT because of the fruitful relationship between the CEO of ITDP, Walter Hook, and Penalosa. Paul Steely White, Executive Director for Transportation Alternatives and formerly the Africa Regional Director for ITDP, recounts the establishment of the partnership between Hook and Penalosa.

> This goes back to a conversation I remember when Enrique was about to step down in Bogotá because his term limit ended and he was angling for a visiting professorship at NYU and he came and sat down on the couch in ITDP's office, which was at the time very small, basically comprised of Walter and myself and Enrique says, "Look, as you know Transmilenio happened in Bogotá and we are good friends of ITDP and Sustainable Transport Magazine…" [41]

In 2003, Penalosa joined the ITDP Board of Directors and together they launched the first BRT tour through South Africa.

Supporters of ITDP attribute its success in BRT distribution, to its constant commitment to BRT advocacy. Many declare that ITDP was the first to bring BRT to South Africa and while most consultants come and go, ITDP's persistence has made a significant difference to the outcome. Part of ITDP's success is attributed to its ability to form trusted networks with local officials [22; 70; 93]. ITDP also makes an effort to involve industry and political officials in BRT discussions [61; 69]. For example, Deputy Minister Jeremy Cronin, who is part of ITDP's trusted circle, recounted how ITDP has been critical both in terms of popularizing BRT among the political and technical stratum, and in creating a platform for communication [62]. In return, Hook credits Cronin for his role in BRT conversations [73]. Many support the involvement of international advocates in the dissemination of widespread practices, so long as they collaborate with South African organizations to experiment with and localize best practice.

Critics of ITDP's approach maintain that the organization leverages a multitude of roles, serving as a salesperson, policy mobilizer and intermediary and thus is too heavily embedded in BRT implementation [33]. The organization is reproached for utilizing its role as expert to be hired as paid consultants to work directly on BRT projects. Ajay Kumar, Senior Transport Economist at the World Bank, plainly states, "ITDP is part of the problem because they try to impose ideas from other countries into South Africa without understanding the local context". This also happened in India where Kumar concludes that BRT was inappropriate for the local context [32]. While ITDP offered capacity to the South African cities to build BRT systems, the extent to which ITDP pushes BRT too hard is still up for debate, mostly because ITDP first created the guidelines for BRT and then commended those cities that followed their instructions. Such valorization of the role of ITDP overlooks the role of local practitioners in BRT flows, which as this chapter demonstrates has been championed by South African policy actors and their associations. Greater understanding of these networks warrants further attention.

Networks of South Africans

Policy tourism, characterized by study tours, site visits and other fact-finding trips, including conferences and even consultancy, is a critical means of distributing best practice across networks of South Africans (Cook and Ward 2011, González 2011). Study tours have become a standard method through which South Africans garner evidence from international contexts, because they enable participants to learn directly from those who implemented the best practice. In pursuit of these lessons, hundreds, if not thousands, of South African political and technical leaders have been to South America to learn of Curitiba's innovative land-use planning, sustainability and transportation solutions, and in the last decade, hundreds of public transportation enthusiasts visited Bogotá to see its thriving Transmilenio BRT (Cameron 2007). By taking policy actors out of their local sociopolitical and geographical context, these adventures overseas connect delegates with one another and with their hosts, with the expectation that partakers will return home and mimic the best practice, thereby forwarding the circulation of policy. This exploration of policy mobilities reconnoiters these "mobility events" and their outcomes, to demonstrate that they are a necessary informal infrastructure through which best practice travels, in particular as a method for strengthening social bonds between delegates and with hosts.

Complex systems have been developed in Bogotá and Curitiba, among others, to manage the throngs of policy tourists and the requisite set of site visits and workbooks needed to promote themselves as role models (González 2011). A host of influential international actors from public transportation advocacy groups and transnational consultants to philanthropic organizations and bus manufacturers lead and finance these trips. ITDP and Embarq are two of the principal sponsors for transportation-related visits to South America. There is criticism of these sponsors who are oft to depict the exemplary projects in a particularly favorable light, often silencing critiques (Wolman 1992), and thus such forms of learning are said to occur only within "ideologically prescribed parameters" (Peck 2011b: 778) in which the participants see, and are shown, the successes divulged previously in publications and talks.

Since 2006, South Africans involved with BRT took part in at least one study tour usually to at least three South American cities. Johannesburg was the first city to participate in a formal study tour to Bogotá to learn about BRT, first in August 2006 and once more in August 2007. Representatives from Cape Town also travelled to South America twice, once in December 2007 and again in December 2008. The first visit invited those technical consultants to learn the details of BRT. As in Johannesburg, Cape Town's second visit was more political. Once local South African systems opened, South African cities visited one another – Rustenburg for example visited BRT systems in Cape Town and Johannesburg in November 2011 instead of traveling to Bogotá, to see how BRT was implemented within South Africa. National politicians also participated in study tours to South America: in 2008, as leader of Parliamentary Portfolio Committee on Transport, Jeremy Cronin organized a study tour to Bogotá, Buenos Aires and Curitiba. It is interesting to note that the National Department of Transport never officially visited South America, an opportunity which might have enabled national officials to provide further guidance on implementation as well as build stronger relationships with city officials. Figure 4.4 details the various BRT study tours.

Not all delegates were as confident that these expensive adventures to Bogotá were essential to BRT approval, as evidenced in remarks in which some wondered how much was actually learned on these trips. One participant articulated, "I think virtually every person even tangentially involved in BRT has been to South America on a study tour. So much money has been spent and so many people have been on joyrides so I would be hesitant to recommend study tours". This skeptic reasons that delegates visit cities implementing BRT because the trip is free for participants, and thus he is concerned that money is wasted sending possibly uninterested parties on "joyrides" all over the world. Perhaps the number of BRT-related study tours has been overstated. Evidence from interviews revealed that 32 percent had been to Bogotá, leaving the remaining 68

South African municipal BRT-related study tours				
City	Date	Destination	Participants	Funders
Cape Town	1st visit: November-December 2007	Guayaquil and Quito, Ecuador; Bogotá and Pereira, Colombia; and Sao Paulo, Brazil	Political officials, technical officials and local consultants	City and delegates
	2nd visit: November - December 2008	Bogota and Pereira, Colombia	Political officials and Paratransit representatives	City from the Public Transport and Infrastructure Systems Grant (PTISG)
eThekwini	Feb-02	Bogotá, Colombia	Political Officials and Paratransit representatives	City
Johannesburg	1st visit: 23 August–1 September 2006	Guayaquil, Ecuador and Bogotá, Colombia	Political Officials and Paratransit representatives	Institute for Transportation and Development policy (ITDP) and the city
	2nd Visit: 17-26 August 2007	Bogotá and Pereira, Colombia	Mayor and 20 taxi industry operators	City from the PTISG
Nelson Mandela Bay	2007	Bogotá, Colombia	Municipal transport officials and paratransit operators	ITDP
Rustenburg	1st visit: November 2011	Cape Town and Johannesburg	Municipal transport officials and paratransit operators	City
	2nd visit: Planned for 2014	Curitiba, Brazil and Bogotá, Colombia	Paratransit operators	City from PTISG
Tshwane	27 February–11 March 2007	Bogotá, Colombia and Paris and Rouen, France	15 city officials, technical staff and paratransit operators	City

Figure 4.4 Details of South African municipal BRT-related study tours.

percent to learn from other sources. Figure 4.5 illustrates the percentage of interview respondents who went to Bogotá.

Unlike other learning methods, study tours provide a more nuanced opportunity through which importing policy actors view the innovation, but also hear about the challenges and failures of the innovation. These international visits enable actors from the same locality to interact outside their ordinary confines. For instance, Johannesburg-based taxi operators, who previously tried to kill one another, found commonality in Bogotá. When they returned, the same operators who until that time refused to work with the city to improve public transportation, were interested in understanding how BRT could work in Johannesburg. Similarly, while in South America, certain taxi operators emerged as leaders and decision-makers, and when they returned to Cape Town, some parlayed this newfound trust into management positions in the new bus company. The remainder of this segment on policy tourism considers the extent to which these exploits are influential in BRT adoption. It unravels the visits undertaken by South African political and technical leaders to see the BRT systems in Bogotá. It also explores the relationships that formed as a result, and evaluates their value in influencing the decision to accept or reject BRT.

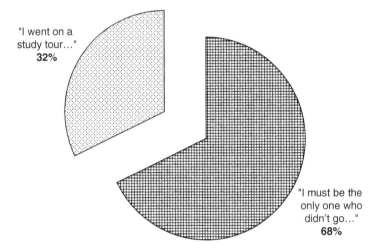

Figure 4.5 Percent of respondents who went on a study tour to Bogotá.

Contrary to popular opinion that all South African involved with BRT implementation travelled to South America, only 32 percent of interview respondents participated in a formal tour, leaving 68 percent of interview respondents without a first-hand experience of BRT from elsewhere.

In August 2006, a team which included Mayoral Committee Member, Rehana Moosajee, Executive Director for Transport, Bob Stanway and Executive Director of 2010, Sibongile Mazibuko as well as representatives from the two local bus companies (Metrobus and Putco) and the two largest minibus taxi associations (Johannesburg Regional Taxi Council and Top Six Taxi Management), visited BRT systems in Bogotá and Guayaquil. The two taxi industry leaders as well as a full-time "tour guide", Lloyd Wright, were sponsored by the ITDP. This initial visit was a "low-key" affair – just a tiny group of officials from Johannesburg accompanying an NGO. The group found their way around Bogotá, met with a few key technical individuals such as Edgar Enrique Sandoval, former managing director of Bogotá's Transmilenio and Darío Hidalgo, Director for Research and Practice at Embarq, another NGO helping cities implement sustainable transportation solutions, both of whom have practical experience building BRT. Because of this initial meeting, Embarq later worked directly with Johannesburg in building Rea Vaya. The study tour provided an opportunity for participants to imagine the possibilities of building BRT in Johannesburg and, perhaps most importantly, marked the beginning of long-lasting and necessary relationships with both technical and political assistants. Upon returning to Johannesburg, the City approved, in principle, Phase 1 of a BRT project. Subsequent tasks included a feasibility study to evaluate the possibility for upgrading the existing SPTN to full BRT; various engagements with the National Department of Transport to finance the project and the Johannesburg Roads Agency and the JDA to renegotiate their contracts regarding the implementation of the SPTN; and initial consultations with the bus and taxi operators affected by the proposed BRT corridor. Two months later, staff returned to the Mayor with a project proposal for BRT (Stanway 2006). All this emerged from a 10-day trip to South America (McCaul 2006a, 2006b).

Johannesburg's second visit in August 2007 to Bogotá and Pereira, included representatives from 17 of the 18 taxi associations potentially affected. The group was quite diverse – prior to this trip, some delegates had never been on an airplane, many did not have a passport and none had seen a bus with capacity for 160 passengers. Obviously, a big part of the study tour was physically experiencing the system, but there were also many critical engagements between informal operators from Johannesburg and their counterparts in Bogotá – they shared experiences working alongside the city in developing BRT, their practices as formal sector employees and even the pitfalls of incorporation. The operators in Pereira told the visitors that they were so opposed to BRT in the beginning, that they blocked the road between Pereira and Bogotá and threw their keys into the river. However, in sharing their initial hesitation, they helped Johannesburg-based taxi operators understand the many ways in which BRT can benefit them. Meeting the operators, even those who initially opposed BRT, helped Joburgers reinterpret their own situation (Stanway 2007).

Study tours are particularly critical in relationship building: before the tour, you could not get representatives from the Johannesburg Regional Taxi Council and Top Six Taxi Management into the same room. Visiting South America created an opportunity to interact outside the usual charged home turf and they realized, disclosed one delegate, that they could not afford to be divided in the face of a strong government eager to establish BRT; if they worked together, as equals, they would be better able to negotiate with the city for their share of the BRT. When they returned, they formed a steering committee and a technical working committee with representation from the 18 affected associations to consult on BRT uptake. One operator defended the study, adding that in Bogotá he realized that his experiences were not exceptional and that if his equivalents in Bogotá could work with their city, perhaps he could at least meet with Johannesburg representatives. Another delegate revealed that through the study tour, the taxi operators developed a rapport in which they were no longer rival operators but South Africans eager to see their country prosper. The initial opportunities for a working relationship between the affected taxi operators and the city emerged from this second study tour.

In November 2007, a delegation travelled from Cape Town to five cities in South America – Guayaquil and Quito, Ecuador; Bogotá and Pereira, Colombia; and Sao Paulo, Brazil – to learn about BRT. The group consisted of two city councilors, three city officials and six private transportation engineers and consultants (who paid for themselves and were currently involved in public transportation initiatives in Cape Town). No representatives from any of the paratransit (bus, minibus taxi or rail) accompanied the group. The official objective of the study tour was to apprise local politicians and technical staff of the range of international practices with BRT systems. By this time, Cape Town's MyCiTi BRT system was already underway, so participants were eager to absorb practical, technical knowledge of assembly, financing and operations (HHO Africa 2007a). In Cape Town, partnerships between local consultants and international experts were a prerequisite for tender. Those traveling to South America did so expecting to build relationships with experienced specialists whose expertise would likely improve the outcome of their project proposal. Consultants that did not partake in the visit had more difficulty ascertaining the necessary international support and were therefore unsuccessful in their submissions. About half of the delegates who travelled to South America in 2007 remain deeply involved in MyCiTi construction and operations, and many attribute the beginning of BRT in Cape Town to this initial study tour.

In Bogotá, the delegation met with a broader array of individuals including the CEO of Transmilenio, the developer of the Transmilenio business model and a fare system expert. They rode the major corridors of Transmilenio to study the experiences aboard the bus, explored the city center to observe the transit-oriented development projects and toured the Transmilenio headquarters to understand how the intelligent transportation systems operate. The visitors also attended a two-day conference on Transmilenio with presentations from local and international transportation experts. In the other cities, the delegation travelled on local BRT systems, reconnoitered the motorized and non-motorized transportation alternatives, and met with the director of urban planning, the director of transportation and the director of the local BRT system. Meeting these experts overseas enabled Cape Town planners to call upon them later.

The official purpose of these visits was to learn about the technical features of construction, financing and operations, as well as the political means to sell BRT public transportation construction and operations, transit-oriented development, systems management and sustainable design from these international cases of best practice, with a view towards implementing BRT in South Africa. More than the material content exchanged, these study tours were necessary for building relationships between South Africans and South Americans as well as between delegates – connections which enabled BRT acceptance across South Africa.

Power Dynamics of Networks

How did these formal and informal associations form and to what extent did this influence the power relations within the circulation of BRT concepts through South Africa? These concerns were debated with some policy actors voicing experiences in which BRT was pushed from the outside [23; 59; 75] and others in which it was dragged in by fellow South Africans [22; 55]; and still others that said it was a combination of both [6; 18]. Andrew Boraine, who rationalizes that both technical and political policy actors were instrumental in mobilizing BRT, summed up this question qualitatively. "Did ITDP start with the technical people and then parachute up to the politicians or did they start with the politicians who instructed the technical people? Well, it was probably a bit of both parachuting together because sometimes these things need to be done that way" [17]. This theoretical approach frames the overall argument regarding the way in which agency is distributed across a variety of personalities, interactions and synergies, reasoning that no individual is capable of mobilizing innovation alone but rather concepts are spread through "translocal assemblages" (McFarlane 2009) composed of a host of local and international, professional and voluntary, government and nongovernment actors and associations.

The pre-eminence of knowledge from outside South Africa and the dominance of international consultants, was described by Pauline Froschauer as her rationale for interpreting BRT as being pushed from the outside. "We were at a stage where anybody who was from somewhere else knew better than we did.

We are particularly bad with that in South Africa ... It is always that if somebody from somewhere else says something then they must be right even if they are wrong but if somebody from here says something, they must be wrong" [75]. This statement suggests that international consultants have the power to ground BRT, albeit geographically and politically detached from the local context. However, as this chapter has revealed, local voices may be fewer in number but hold substantially more power to act on published information.

By contrast, a number of South African respondents claimed that a mixture of international and local advocates fueled the adoption of BRT. It was an instance of "us going out and learning elsewhere and applying it to our South African context", concludes Seana Nkhahle, Acting Executive Director of Strategy, Policy and Research at the South African Local Government Association (SALGA). He reasons that before learning from the outside, South African actors identified the possible transportation solutions and then sought information on how to implement them [6]. Rehana Moosajee, like Nkhahle, supports policy exchange from others but asserts, "We have to stop apologizing for learning from others because if we are not willing to learn then how is the country going to grow" [22]. These contentions are not intended to exaggerate the role of local actors, but rather indicate that if the policy mobilizers had not bought it in, BRT may have still filtered into South Africa through some other means. "If a new water purification system is developed in Japan today, then within months, they will be bringing it here", suggests Christoph Krogscheepers, Director at ITS, the company responsible for designing the intelligent transportation system components of MyCiTi in Cape Town. "Even if Lloyd never came to the country, our own engineers would have started looking at it and going there" [61]. However, "once we invited it in", rationalizes Froschauer, "it came full pressure, like a water tap" [75].

The simultaneous pushing-and-pulling within these relationships exposes the uneven power distribution across these relationships. While international advocates are instrumental in introducing the innovation, local intermediaries, politicians and pioneers are responsible for localizing the concept. Even Lloyd Wright acknowledges these limitations when introducing BRT to new cities clarifying, "I don't think any project will ever succeed if it's being internationally-driven. ... It has to be locally-driven and locally-owned. ... There is a distinct international and local role but policy and project development has to be owned locally". Wright's claims that policy exchange is not about innovative technology or impressive planning, but about maneuvering and utilizing interpersonal relationships to get local support for the project. This is one of his greatest lessons from South Africa – and summed up well by Bob Stanway, "It is almost by chance that one does get a few people together that can do something and that complement one another and have a shared vision. It is quite difficult to do that" [27].

Conclusion

This chapter explored the actors and associations moving BRT across South African cities. It revealed that best practice policy models are mobilized by a network of global and local policy actors including policy mobilizers, intermediaries and local pioneers. In this instance of BRT adoption, South Africans enticed BRT in through their support of international consultants, but likewise, it was pushed from the outside by charismatic proponents teaming up with international policy networks to enthuse South African implementers. Often interview respondents attributed BRT to a particular proponent but, upon further reflection, agreed that no single individual can be attributed with responsibility for moving BRT policy across South African cities. Therefore, it was within and through pre-existing associations between global BRT advocates and South African governmental and nongovernmental actors that BRT was undertaken. While strong personalities were at times dominant, this was balanced by the power structures within South Africa, which endowed local implementers with the power and responsibility for localizing the BRT model. Whereas international policy actors may have possessed greater knowledge and experience with the model, the actualization of the model took place within South African cities through local policy actors.

The various narratives introduced above offer a critical contribution to the literature, by addressing the way in which learning takes place, the role of various policy experts and the power structures employed to adopt policies from elsewhere. These arguments add a critical contribution to the policy mobilities field by suggesting that a variety of policy actors are instrumental in assembling, mobilizing and adopting mobile forms of knowledge. Not only does this chapter provide a strong empirical case of the involvement of human subjectivity in the process of adoption, but it also offers an occasion to reconnoiter the significance of power relations and personality in advancing BRT. I argue that BRT was not adopted simply because Lloyd Wright spoke at that 2006 BRT workshop or after he met with Cape Town's Mayor Helen Zille, but because of the decisions and determinations of local policy actors eager to actualize transportation improvements in South African cities. These local actors need not be employed by the adopting locality, but they understand the particular political constraints and opportunities. This reminds us about the difficulty of implementing circulated policies, and suggests that local politics might play an underexplored role in policy adoption processes. The next chapter situates the presentations by Lloyd Wright within a wider understanding of the South African city, to uncover how these policymakers made the particular decisions they did regarding BRT knowledge.

Chapter Five
The Local Politics of BRT

Introduction

The previous chapters presented BRT adoption as an orderly process through which South African policy actors learned of and implemented the innovation. Whereas Chapter 4 focused on the activities of policy experts and international consultants boosting the best practices of select sites and drumming up enthusiasm for BRT, and Chapter 3 explored the process of policy mutation taking place along the way, neither attended sufficiently to the local landscape into which policies arrive. This chapter reveals that support for BRT was not an inevitable progression in which policymakers learned of the innovation and gradually adopted it. While each South African city was exposed to the BRT model by similar policy mobilizers at roughly the same time, each city interpreted and applied it in accordance with the local government considerations and needs. Whereas policymakers in Johannesburg quickly adopted BRT, followed immediately by their counterparts in Tshwane and Nelson Mandela Bay, Cape Town and eThekwini were slower to make a decision. The decision to implement BRT in Johannesburg resulted in a successfully operating system, but in Tshwane and Nelson Mandela Bay there were significant political and technical difficulties that delayed BRT. This chapter examines the international, national and local politics shaping BRT circulation and adoption, in particular thinking about the differences in the decisions made across South African cities.

This chapter explores the politics of policy mobilities. It uncovers the international, national and municipal relationships that facilitate and/or impede

How Cities Learn: Tracing Bus Rapid Transit in South Africa, First Edition. Astrid Wood.
© 2022 Royal Geographical Society (with the Institute of British Geographers). Published 2022 by John Wiley & Sons Ltd.

the opportunity for policy exchange and adoption. Policies are constantly in-motion with some innovations taking root while others remain untethered to any particular place. Policy adoption is uneven and unpredictable with some cities better suited to managing the process, others gravitating towards a particular innovation, and finally there are those choosing not to employ mobile policies. In South Africa, policymakers adopted BRT because of existing conditions, which include both political interactions between political parties and between cities, as well as the availability of technical expertise and limited financial opportunities. This chapter demonstrates how the political context influences policy adoption. It also considers how policy exchange influences, and is influenced by, the relationships between municipalities.

The next section presents the international context of learning highlighting the importance of south–south politics for South African policymakers. It identifies the multiplicity of cities, not just Bogotá, from which BRT best practice disseminates. The discussion focuses on the process for selecting sites of learning and the relationships forming between localities as a result. The second section considers the role of interurban mobilities, or the South African municipal interactions and alignments, in the determination to adopt transferred concepts. The third empirical section looks at the conditions within South African municipalities that influenced the adoption of BRT. Finally, I conclude by reflecting on the role of local politics in grounding mobile knowledge. Such debates deepen and widen the space through which policy flows, by demonstrating how local municipal and regional relationships shape the circulation and adoption process.

The International Context of BRT Circulation

"What is it about BRT that makes it such a popular choice for so many (South African) cities?", I asked a Johannesburg-based development specialist involved with implementing the city's new Rea Vaya service. "What I like about BRT is that it is a south-south connection", she reasoned, "From what I understand, the conditions in Bogotá are quite similar to those in Johannesburg" [36]. BRT dissemination is applauded across the world and specifically in South African cities, for being a southern innovation better suited for cities with fewer financial and institutional resources, and the learning is frequently framed as an instance of exchange between cities of the global south. "It was extremely powerful to say that this is an approach to planning for transportation that has worked in a city that has a lot in common with your city", remarked an international transportation advocate, "more than New York or Houston or Amsterdam or Copenhagen so that south-south transfer was really powerful" [41].

BRT is thus frequently seen as an instance of south–south exchange, with implementers making extensive reference to the cities in South America, but maintaining a curious disregard for transportation solutions elsewhere. This section presents the South African stories and experiences of learning from South America as well as those that are rarely discussed in regards to exchange (and the retention) of knowledge between Lagos, Nigeria, Ahmedabad, India and cities of the global north, whose experiences with BRT profoundly influenced South African adoption and implementation. In so doing, it addresses the shifting geographies of learning by focusing on the documentation, deliberation and discussions exchanged between South African politicians, officials and consultants and their counterparts in Bogotá and elsewhere in South America, to reconsider the exchanges that takes place between cities of the global south. While BRT dissemination has been lauded as evidence of the magnitude of southern engagements, my research interrogates this notion to reconsider the influence of existing relationships between these sites to justify ongoing endorsement in South Africa.

Learning from South America

South African learning from South America has been going on for some time now: South Africa's Truth and Reconciliation Committee was instituted on the guidance from the governments of Argentina and Chile; South Africa and Brazil exchanged experiences related to land reform; and the South African housing subsidy program was modeled broadly on the Chilean model (Gilbert and Crankshaw 1999). None of this has been as widespread or efficient as the ongoing borrowing of BRT, whose popularity is derived from its associations with Bogotá. South Africans indicated a number of reasons for replicating Transmilenio. Several South Africans suggested that it had to do with perceived political and institutional similarities between Bogotá and South Africa – if BRT worked in a city like Bogotá, a similarly sized city with comparable institutional challenges, then it seemed likely it would be as appropriate in South Africa.

Part of the success of the BRT model rests on the rebranding of Bogotá, which transitioned itself from a city once distinguished by its drug trafficking and crime, to one of the cleanest, safest, greenest, and best managed cities in the global south. Perhaps South Africans were averse to publicizing learning from other cities in South America – replicating the successes of smaller cities with little global profile might be more difficult to gain support for within South Africa. It is also likely that Bogotá was exalted because of the political considerations of individual actors, eager to align with the economic and political support for Bogotá to further their own professional pursuits. Regardless of the exact motivation, there is little evidence that South Africans replicated Transmilenio because of any specific technical or operational achievements. The understanding of BRT as a Bogotá-based innovation and an example of south–south collaboration, came directly from South African policymakers: they believed that the South African context was similar to the conditions in Bogotá, and because of these resemblances, BRT would be successful in South Africa. For the most part, South Africans skipped the usual due-diligence and skepticism of international projects because people liked the idea of a south–south connection, assuming universality across cities of the global south. Jeremy Cronin, Deputy Minister of the National Department of Transport from 2009 to 2012 and a staunch advocate of BRT, admitted that South Africa made a mistake by focusing its attention towards learning only from Bogotá. Cronin is sensitive to the local challenges emerging from the "political and spatial complexities of our cities" which Bogotános did not understand "and why should [South Africans] expect them to" [62]? South African cities are far more spatially and socially fragmented, with higher unemployment figures and a more inclusive decision-making process than South American cities. A transportation representative from the World Bank, who sharply criticized South Africa's admiration for full-feature BRT (as opposed to the BRT-lite system first implemented in Lagos in 2007), argues, "Latin American cities are very different from South African cities. In South Africa, the biggest problems are access and affordability. Latin America has different problems" [32]. In addition, historically-speaking South African cities have been served by commuter rail lines, whereas South American cities always relied on road-based systems. This led several South Africans to question such extensive investment in new bus networks rather than upgrading the existing rail network. BRT was developed in a very specific, economic, social, political and cultural context, and "those preconditions are simply not there in many South African cities" [67].

Officials with the eThekwini Transport Authority recognized the vast geographic and sociopolitical differences between South American and South African cities, and therefore postponed adopting BRT. In 2002, Logan Moodley, Deputy Head of Strategic Transport Planning at the eThekwini Transport Authority since 1986, participated in an early study tour to Bogotá. The trip was part of an early foray by the Institute for Transportation and Development Policy (ITDP), who sponsored Moodley's visit. He recounted his impressions of Bogotá, recalling that while he noticed there were similarities, especially in terms of the city's strained relationship with unregulated transportation operators and their propensity for violence, he also observed a number of differences including the variation in population, density, employment and income. He returned from the study tour with reservations regarding the applicability of Bogotá's BRT system in eThekwini, which ultimately led the city to postpone introducing the concept locally [47]. It is important to note that this visit took place in the early phase of the translation of Bogotá model and the sociopolitical power of the model had not yet developed. South Africans were still more familiar with the systems in Curitiba and Quito with only a vague notion of the possibilities for aligning with Bogotá.

Although South African policy actors identified Bogotá as their greatest inspiration for adopting and implementing BRT, there was ample evidence of the successes achieved by Bogotá's forerunners. The original concept of road-based public transportation was developed by Lima's introduction of fully segregated lanes and perfected by Curitiba's architecturally distinct stations and integrated ticketing system. What seemed like best practice pioneered in Bogotá was actually the culmination of almost three decades of incremental improvements in cities across South America (Mejía-Dugand et al. 2013). There is evidence of Curitiba-based planners and politicians visiting the system in Lima to study system characteristics, and later of Bogotá-based experts traveling to Curitiba (Ardila-Gomez 2004; Wright 2007a). In both cases, they returned home and implemented a version that aligned with local demographics and demand. "It is interesting to remember the importance of Curitiba", recalled Rashid Seedat, Head of the Gauteng Planning Commission, "most people forget that Curitiba was the city that started BRT because Bogotá was the city that made it big international news" [40]. Although these earlier achievements did not attract international attention, they certainly had a role in the learning. The question is, in what ways were their successes used, perhaps unofficially, as sites of learning?

In several internal documents discussing the rationale for selecting BRT in Johannesburg, local policy actors identified Bogotá as their greatest influence but also recognized the role of other South American cities in disseminating information and experiences. In particular, they make mention of lessons acquired from MetroVieja in Guayaquil and Metrobus in Pereira, suggesting that seeing these systems ultimately enhanced their understanding of Bogotá. The evidence suggests that policy models like BRT have multiple meanings to implementers and likewise implementers hear from multiple examples. Whereas MetroVieja in Guayaquil inspired Bob Stanway, Project Manager for Rea Vaya Phase 1A [8], Eddie Chinnappen, Executive Director of Transport in the City of Cape Town between 2006 and 2009, was enthused by the complexity of Transmilenio in Bogotá [89]. Eric Motshwane, former chairperson of Greater Johannesburg Regional Taxi Association and now Director at PioTrans, the operating company for Johannesburg's Rea Vaya Phase 1A, recalled his experiences on a study tour as vital in building an understanding of the possibility for BRT in Johannesburg,

> When I went to Bogotá for the first time in 2007, I thought that is not doable. That thing is huge. There are six lanes running that way and another six moving that way. Bogotá is six-times the size of Johannesburg and I thought this BRT project cannot happen here... Only when I went to the city of Pereira and saw their BRT that I realized this can be done in Johannesburg. Their system really inspired me. I also went to Guayaquil but it didn't convince me. We also went to Quito but it had gone down. When I went to Pereira in 2008, in my heart, I believed it could be done. The landscape of Johannesburg is more like Pereira. The trunk is more like ours... Pereira convinced me that Johannesburg could do it.

Motshwane was not interested in replicating a Bogotá-styled BRT system because it seemed inappropriate for the South African context. However, when he saw Metrobus in Pereira, he realized the possibilities for BRT in Johannesburg [94]. These other examples informed the Bogotá model and circulated alongside it.

Sometimes actors discover through constructive engagements, other times by dismissing the experiences of a particular site: Quito's El Trole, an electric trolleybus system, is one such system often overlooked in South Africa. A variation of El Trole was first proposed in Cape Town in the early 2000s when the municipality briefly experimented with a BRT system along Klipfontein Road in the southwest part of the city. At that time, the Quito model was considered appropriate in South American cities, which tend to have a compact city center with narrow streets and a historic urban landscape (Wright 2002). The Bogotá model later replaced Quito, as it was presumed ideal for a more modern, suburban landscape with wide avenues suited for busways in the median and sufficient road space to maneuver larger buses. Pilo Williumsen, a transportation engineer with experience on a number of different BRT systems, including the operational design of Transmilenio and the modeling for Cape Town's proposed Klipfontein Corridor, thought that a variation of El Trole may have been more appropriate for Cape Town and Johannesburg's historic city centers with narrow streets and small city blocks; with the Bogotá model being more suitable for the suburban areas of those cities [51]. Concerns for the technical feasibility of the Quito model were the official justifications for its elimination as a model, but I want to suggest that the representational power of the Bogotá model was a necessary component for South African appropriation. It was more than simply a south–south connection but a particular connection between Cape Town and Bogotá and Johannesburg and Bogotá, that facilitated the exchanges that justified appropriation.

These examples reveal that BRT circulation is more complicated than a rapid, lateral south–south exchange between cities eager and receptive to mobile knowledge. Rather, the process of exchange between cities is asymmetrical, uneven and incredibly partisan. Part of the success of BRT dissemination has rested on the rebranding of Bogotá, which transitioned itself from a city once synonymous with deficient governance to one of the world's dynamic, industrious cities. These accomplishments are assumed to accompany BRT implementation or at the very least be part of a process of transformation. South African cities were eager to replicate this loftier urban transformation.

Learning from Africa

Much of the same rhetoric used to generate support for BRT through the assumed potency of south–south exchanges, was exploited to excuse the minimal learning and exchange across African cities. In spite of South Africa's tendency to exceptionalize its apartheid history, many former colonial cities are characterized by common colonial planning practices – a grid road layout, transportation services that radiate from the city center, and a history of racial and social segregation (King 1976; Lemon 1991). The general sentiment in South Africa, however, is that it "is just a step above the rest of the continent" and thus there is little to gain from visiting them [76]. Such an interpretation associates learning with value – there is understood to be little economic, political, social benefit to building relationships across African cities. Certainly, the strongest policy informants came from South America but to what extent were South Africans also interested in comprehending what was going on elsewhere, in particular their continental neighbors? This is particularly important since so much of the learning surrounding BRT is rationalized as an outcome of geographic propinquity. Walter Hook of ITDP for instance frames the distribution of BRT in this light – "Colombia of course was replicated fastest because they're all right next to Transmilenio. So after Bogotá we saw BRTs open in Medellin, Cali, Barranquilla, Bucaramanga and Cartagena…" (Hook 2014).

A BRT system has been operating in Lagos since 2008. The 22-km corridor moves along a busy, multi-lane expressway with 26 bus shelters and 3 bus terminals. BRT-lite, as it has been dubbed, unlike the Bogotá model does not have electronic fare collection or separate lanes and runs low-floor buses with front door loading, but carries as many riders as Transmilenio and operates without any subsidies. BRT-lite utilizes less infrastructure and therefore is cheaper to build and easier to operate than a full-feature system. Lagos officials focused on organizing the taxi industry rather than building expensive, iconic infrastructure, and thus it is a relatively inexpensive scheme as compared to those in South African cities [32]. Some in South Africa say that Lagos "did a good job" doing "what was possible because they didn't have more money or the political will or the capacity to do something better and it is better to do something that is feasible than do nothing and wait to do something perfect" [51]. Others say that Lagos "should have gone for a full-BRT solution", rationalizing that you only get so many opportunities to transform the urban public transportation network [49].

Here was an African version of BRT, and yet in spite of the aforementioned support for BRT as a southern innovation and an opportunity for south–south collaboration, no study tours went to Lagos. South African interview respondents rarely mentioned the operational BRT system in Lagos or other BRT projects in Accra, Dakar, Dar es Salaam or Nairobi. Often, I introduced the example in Lagos and asked if there were exchanges between South Africa and other African cities. Reactions ranged from "No, we didn't" [8] to "It's a joke and they shouldn't even talk about it" [16]. Professor Philip Harrison, the South African Research Chair in Spatial Analysis and City Planning at the University of the Witwatersrand in Johannesburg and Executive Director of Development Planning and Urban Management in the City of Johannesburg from 2006 to 2009, reveals that South Africans were aware of the Lagos model of BRT-lite but it paled in comparison to "the strength of the model from Latin America". It was not a specific attempt to ignore Lagos, he asserts, but because South African cities had superior financial and technical capacity, there was simply more to gain from elsewhere [4]. There was also a mention that Lagos was disregarded because South African cities "went for the gold standard" with cities proudly building "pure BRT" [5].

Other continental neighbors, namely, Accra, Dakar, Dar es Salaam and Nairobi, faced similar challenges in implementing a BRT, which might have been informative for South African cities. In Dar es Salaam, for instance, the largest city in Tanzania and one of the fastest growing metropolitan regions on the African continent, it took almost 15 years before launching a BRT in 2015. According to the city's Transport Policy and System Development Master Plan, Phase 1 runs from Kimara to Ubungo and includes 21 km of dedicated lanes running in a closed system, 29 stations and 5 terminals at a cost of USD125 million. The project was initially introduced to alleviate the city's worsening traffic congestion and improve the movement of passengers through the city. Policymakers in Dar es Salaam, like many of their counterparts in South Africa, learned of BRT through ITDP and their 2002 "Building a New City Tour" (see the 'Forming the Bogotá Model of BRT' section). In Dar es Salaam, however, implementation has been especially cumbersome (see Figure 5.1 for the status of construction in 2012). The project is limited by inner-city politics; because Dar es Salaam is composed of three municipalities and a metropolitan government, responsibility for the project is

Figure 5.1 Shekilango BRT Station, Dar es Salaam.
This is an incomplete BRT station on the corner of Morogoro and Shekilango Roads in Kinondoni, near the main intercity bus terminal on Route 112X, the 10.8-km express path from Ubongo to Kivukoni (Photograph by author April 2012).

divided across four town planners, four municipal directors and four mayors. Moreover, the municipality is struggling to organize the minibus taxi operators who, like in South Africa, were offered the opportunity to become operators of the new system. There are also challenges in the expropriation of land along the BRT route.

The point here is not a retaliatory exposé of who was ignored by whom, but to consider the wider politics of knowledge exchange. Evidence suggests that there was a specific effort to disregard the merits of BRT-lite because South African cities wanted to emulate the most advanced examples of pure BRT elsewhere. This demonstrates that global south cities are not inherently more likely to learn from each other but rather policy mobilities is like other elements of policymaking in which politics are preeminent. Furthermore, it suggests that geographic proximity is not the most important indicator of exchange; rather, a willingness to learn, perhaps generated by grander intergovernmental narratives, is what drives policy adoption.

Learning from India

It is also particularly surprising that there was so little engagement with Indian cities concurrently building new public transportation systems. According to some informants, there was insufficient engagement with Ahmedabad whose BRT is renowned for its low-cost buses: Ahmedabad's Janmarg (meaning, people's way in Gujarati) was initially proposed in 2003 by the state as a substitute for a rail-based metro system, which it considered too expensive. The following year, a local planning institute invited Enrique Penalosa, former Mayor of Bogotá, to present his experiences building Transmilenio. In 2005, the government initiated a feasibility study and, in 2006, the Ministry of Urban Development approved construction of the first 12 km of a proposed 53 km of Phase 1 for a total of 217 km of BRT corridors (Kumar et al. 2012; Mahadevia et al. 2013). Janmarg is remarkably similar to Johannesburg's Rea Vaya – both cities constructed 900 mm high-floor stations and operate environmental-friendly buses, and passengers pay their fares before boarding through a cashless system. Moreover, the cities adopted and implemented BRT concurrently, gaining technical and financial details through the same international platform – Enrique Penalosa and ITDP – and yet there was minimal, if any, engagement between the cities.

There is certainly a strong connection between South African and Indian cities, not least through the millions of South Africans of Indian descent whose cultural and financial exchanges across the Indian Ocean are noteworthy (Hansen 2012); however, neither nation sufficiently exploited these relationships. Experts from India came to Johannesburg in June 2007 to share their experiences with building density through transit-oriented development. While "the strongest policy informant was from Latin America…there was also an awareness of what was going on elsewhere", discloses Philip Harrison [4]. There were also instances when South Africans travelled to India to learn of their best practice. For instance, the Housing Department in Johannesburg went to Mumbai to see their land adjustment schemes but they did not like the model – to hand over land to a developer and develop a portion of the land for high-income real estate – so they did not implement it. One international transportation specialist expressed stark concern for the lack of engagement with India since he believes that South African cities have more in common with the higher concentration of poverty and sprawling landscape of Indian cities as opposed to South American cities [33]. Some South Africans, however, regarded Indian cities as "too chaotic" and their urban challenges such as providing clean water and proper sanitation "seemed worse", so local actors "couldn't really get excited about the Indian projects" [4]. Perhaps South African cities were cautious against elevating the interventions in Indian cities whose position as a key player in the global south threatens South Africa's role as a regional leader. This raises an interesting question: is it possible to have too much engagement? Is there a downside to south–south connections? These questions will be explored further in the following section examining learning from the global north.

Learning from the North

The discussion of Indian cities suggests that South African cities are still captivated by learning from the global north and thus they were drawn to BRT because of their associations with cities like New York and Copenhagen. Zahira Asmal, editor of Designing South Africa (2012), a collection sharing the experiences of South African experience as host of the 2010 Football World Cup with Brazilian cities, hosts of the 2014 Football World Cup summed it up best: "We talk about south-south dialogue but we always prefer north" [85]. Gil Penalosa, Executive Director of 8–80 Cities, an NGO based in Toronto, Canada, repeated this logic when he came to Johannesburg in May 2012. As part of a weeklong engagement with the city, Penalosa gave a public lecture, "Transforming our Streets: Bringing the International Experience to Johannesburg" (2012), in which he presented more than 600 images of various urban interventions around the globe, many of which showed projects in Amsterdam, Copenhagen and New York City as well as Bogotá and Mexico City. After the talk, I asked Penalosa why he showed so many examples from the global north given the differences between those cities and Johannesburg. He defended the importance of presenting examples from both affluent and modest cities to show the pervasiveness of the model, presuming that many South Africans see themselves as part of these more affluent cities and are therefore drawn to replicate those transportation interventions [79].

This evaluation helps us consider how and why certain cities are brought into conversation with one another and what happens as a result. In this instance, South African cities aligned with Bogotá because of broader considerations for achieving urban redevelopment and climbing global hierarchies. The Lagos model of BRT-lite, by contrast, was disregarded not because of any financial or institutional inadequacies but because of politics, which privileged a variation elsewhere. It need not have been Bogotá which became the paramount model, however. It could have been Quito if the leaders from Quito had formed relationships with these municipalities to elevate their achievements over those taking place in other cities. These conditions are perhaps even more stark in cities where former colonial relationships still exercise power in local policymaking.

In explaining the ways in which cities circulate knowledge, this section highlights the importance of south–south connections in local policymaking. These arguments make an important contribution to the existing literature on policy mobilities by mapping and explaining the uneven geographies of south–south policy exchange.

The National Context of BRT Circulation

Competitive tendencies are one possible explanation as to why South African cities continue to adopt and implement BRT, even though its local viability remains dubious. "No city wants to be the only one that doesn't have a BRT system", deduced one transportation analyst at the World Bank. Recognizing the peer pressure South African cities sense in the adoption of BRT, he stated that "cities are not necessary responding to a demand in the transport sector but considering the actions of other cities" [32]. Analogous anxieties for the rapport between cities implementing BRT were repeated by an array of South African politicians and planners, pointing to an often overlooked association between policy exchange and interurban competition. "Competition between cities in the same region is a race to the bottom", articulated another South African policymaker concerned by interurban relationships, "Cities need to be competitive but not in competition" [17]. This part of the chapter attends to the process of BRT circulation as it swept through South African cities, to reconsider the role of local municipal interactions and alignments in the determination to adopt best practice.

Rivalry flourished among provinces and their municipalities, as well as across municipal agencies, for control of the transportation function [31; 83]; affected minibus taxi operators also battled for their stake in BRT profits [92]. For some, the transfer reified the unspoken hierarchy between the South African cities – Cape Town and Johannesburg being a step above the rest. No city wanted to be the "guinea pig" leading one transportation engineer to express gratitude to Johannesburg and Cape Town for being first, rationalizing that because their BRT systems will need to be subsidized, "there has been a sobering up" and a better understanding of how to calculate ridership figures and financing [49]. There is competition for limited funds from the Public Transport Infrastructure and Systems Grant (PTISG) – a conditional grant from National Treasury available to most South African cities that apply – and as more cities begin constructing systems, this antagonism will increase (see the "State Intervention in Transportation" section). There is also a battle for limited technical capacity, since there are only so many consultants both in South Africa and internationally capable of designing and building BRT. The rivalry heightened by the PTISG might explain why so many South African cities have adopted BRT, even though both construction and operational costs are higher than expected. Concerns such as these are practical given the challenges with localizing an international policy model, and raise theoretical interest in understanding the way in which interurban relationships and interactions boost, deter or have insignificant influence on adoption practices.

The case of BRT circulation in South Africa offers an opportunity to consider the influence of competitive relationships on policy adoption. The intellectual challenge here is to reflect on how cities can boost their achievements without heightening local competitive tensions, and as such compete without being competitive. Certainly, the reduction in state staffing budgets coupled with intensified competitive pressure and shortened deadlines for financing, might increase the attractiveness of adopting policies from elsewhere; but, undoubtedly, these processes demand further unpacking under the lens of interurban mobilities. The evidence below suggests that South African cities adopted BRT because of extant sociopolitical conditions, which contain both political interactions, including contestation between political parties and concerns for internal hierarchies, as well as technical circumstances, involving the availability of technical expertise and limited financial opportunities.

Political Interactions between South African Localities

Competition between South African politicians was understandable in the lead up to Rea Vaya opening in August 2009. "There was a big fuss about Johannesburg launching the BRT first," mentioned one Capetonian, because every South African city wanted to be the first to launch their system [12]. Richard Gordge, Director at Transport Futures, offers an alternative understanding of South African municipal interactions and relationships, explaining that when Rea Vaya started operating in 2009, South African cities were at very different phases. "Cape Town came to the party very late; Johannesburg had Rea Vaya; Nelson Mandela Bay was a complete mess with the taxis; Durban was reluctantly married to its rail strategy, so it didn't get out of the box very quickly; and Tshwane was all over the place. So nobody knew where to look at in South Africa" [64]. Certainly, cities in the same region implementing BRT simultaneously should interact and yet as Gordge sees it, these cities did not sufficiently exploit the opportunity to collaborate.

It is important to account for the South African political conditions guiding interurban mobilities: the African National Congress (ANC) is the dominant political party in the national government, supported by the tripartite alliance with Congress of South African Trade Unions (COSATU) and the South African Communist Party (SACP). Until 2016, the ANC governed all cities except Cape Town, which to date is still managed by the Democratic Alliance (DA) along with its surrounding Province, the Western Cape. The Inkatha Freedom Party (IFP) is the fourth largest political party with strongholds in eThekwini and KwaZulu-Natal. Rashid Seedat, Head of the Gauteng Planning Commission, illustrates the influence of politics on the interactions even among ANC-controlled cities: "There has been a competitive relationship between the metros and the provinces and between the metros and themselves on everything – like a lack of cooperation, not a hostile competition because they are all in the same political party, but there was never a sense that they needed to cooperate" [40]. National politics spills into local government administration, particularly in Cape Town and Johannesburg where opposition between the ANC and the DA is manifested in the relationship between the two cities. Politicians, however, deny any rivalry. Baleka Mbete, the National Chairperson of the ANC since 2007, rejected any conflict between Cape Town and Johannesburg, when I asked her if politics was a factor in BRT support. "The two cities are two hours away from each other [by plane], of course they work together" [15]. Michael Sutcliffe, City Manager of eThekwini from 2002 to 2011, was quick to add that these are just "political games" because each city has "its own profile and its own challenges" so there is "no deliberate, conscious strategy" to compete [78].

The most partisan illustration of intercity interactions is evident in the exchanges between politicians in Cape Town and Johannesburg, who supposedly bridged their differences as political rivals (Cape Town being controlled by the DA and Johannesburg being governed by the ANC) and as competing economic interests (both cities compete for economic power in South Africa) to advance BRT adoption. The political leaders of transportation in Cape Town and Johannesburg, Brett Herron and Rehana Moosajee, rebuffed the suggestion that any competition occurred between the cities, each praising the other in interviews. In 2009, shortly after Rea Vaya began operating, officials, including Elizabeth Thompson, then the political head of Transport in Cape Town, came to meet with Johannesburg's transportation team and ride the new system. Mayor Amos Masondo of Johannesburg and Mayor Helen Zille of Cape Town decided to sidestep politics and share their local experiences with BRT. Delegates from Cape Town returned twice more to review how Johannesburg had implemented the Bogotá model. Moosajee and Thompson met on several occasions in which the two discussed issues around construction, taxi transformation and financing. Nkosinathi Manzana, former Chief Operations Officer at the JDA with overall responsibility for the BRT portfolio, remembers that "in the later stages, there was a bit of competition going on" but "Cape Town asked and Johannesburg opened its doors and Cape Town came in and had a look and said this and that and we gave our pointers of certain things to avoid" [29]. Herron similarly praises Moosajee clarifying, "We agree that our priority is improving public transport and not competing...There is a lot we can learn from Johannesburg'. Herron continues, 'That is why I'm going to see what they've done and how their operations work and where they are going and I hope that we can share our successes and our challenges" [65]. Such a good rapport between political rivals leaves me wondering if these interactions are honest, substantive, useful exchanges of both achievements and disappointments, or just another round of political games. Moosajee responded to my doubts illuminating, "there was quite a bit of collaboration between cities" but "it could have been more" and "my own sense is that we haven't had enough *in-depth sharing*" (my emphasis). She rationalized that sometimes sharing comes across "as if people just want to blow their own trumpets and talk about the difficulties and talk about how it was for them but we are somewhere else in our journey" [22].

Part of the problem with these forms of collaboration is peer pressure. Every city wants to have the paramount BRT system, observed Ajay Kumar, Senior Transport Economist at the World Bank, at a workshop in Nelson Mandela Bay in November 2011, to help the city consider how best to proceed with its stalled BRT project. Kumar added, "cities are not necessarily responding to a demand in the transportation industry but considering the actions of other cities" [32]. This peer pressure easily morphs into rivalry as evidenced by Maddie Mazaza, who in telling me about her visit to Johannesburg in 2009 revealed the inherent rivalry between Cape Town and Johannesburg. "There was a lot of interaction during that time", she reckoned, "but you see, Johannesburg doesn't have a plan for non-motorized transport and we said we wanted non-motorized transport. Johannesburg doesn't have a control center and we wanted a control center... We are leading now on the MyConnect chip card but Johannesburg doesn't have this yet" [70].

Nowhere are these interurban rivalries stronger than in Gauteng Province, where the three largest municipalities, Johannesburg, Tshwane and Ekurhuleni, each want to appear as the most proficient and economical manager of their transportation sector. These competitive tendencies led to so many "missed opportunities" for collaboration and exchange between these neighboring municipalities [88]. One such opportunity took place in April 2012, when representatives from Ekurhuleni (who at that time were still only considering adopting BRT but in November 2012 approved plans to proceed with three lines) visited Eddie Chinnappen of Cape Town's MyCiTi to hear about his experiences with BRT. Chinnappen admitted that it was "strange for Ekurhuleni to come and see me" because he was no longer employed by the City of Cape Town [89]. Instead, they should be building strategic ties with the cities operating BRT, not just those actors involved with a particular route. It might have been more advantageous for these representatives to have used BRT to build relationships with neighboring metropolitan municipalities in Gauteng, both of which had extensive experience building BRT in South Africa.

One outcome of this neglectful approach to learning from other cities, is the decision by officials at the National Department of Transport to require the subsequent lines of BRT be designed for low-floor buses and platforms. Jonathan Manning, the chief architect of the Rea Vaya stations, assumed that the shift from high-floor to low-floor buses means that the new systems in Ekurhuleni and Tshwane will be incompatible with the operational system in Johannesburg [34]. This physical incompatibility in the contiguous municipalities of Gauteng, is further illustrative of the political indifference between the three cities concurrently adopting BRT.

Technical Exchanges between South Africa Localities

More than simply magnifying the condition in which "every city wants a BRT bus and ignores their own needs rather than just improving their current bus system" [21], a deeper analysis suggests that intercity competition is heightened through the technical exchanges taking place through the adoption of BRT. Many were concerned that localities were not learning from one another because of the competitive atmosphere associated with major infrastructure investments. One Capetonian consultant was bothered that "the scope for interaction and cooperation between Cape Town and Johannesburg has been amiss" and while there may be interest in building these relationships, the demands of implementation prevent substantive exchange [54]. Another expressed concern that the opportunities to collaborate were "right in front of them and most of the cities didn't grab it" [64]. Andrew Boraine describes competition between South African cities as a race to the bottom. "Cape Town's main competitors may be Toronto, Singapore and Buenos Aires for argument's sake, and Johannesburg's could be a set of other cities and the futility of arriving in one part of the world one month and talking down your competitors and they arrive there and do the same to you. Everyone is just cutting their own throats". He senses that Cape Town has more in common with Johannesburg and thus greater motivation to cooperate rather than compete [17].

Perhaps most pressing was the competition for funds from the PTISG, which became a critical source of contestation since all cities were bidding for the same stream to fund BRT infrastructure (see the 'State Intervention in Transportation' section). "One of the biggest problems ...and all cities will deny it ... is that there is huge competition between the cities because there is money involved", deduced Pauline Froschauer, Director of Namela and managing director of the implementation of BRT in Rustenburg. "All the money is held in the same pot" so "there is competition between cities for the same money". Funds from the PTISG are divided between those cities implementing BRT and "there is just no way that there is enough for all the cities". Each year, cities present their BRT plans to the National Department of Transport and National Treasury and request funding for the next three years [75]. Policy actors complain of the "use-it-or-lose-it approach" which forces cities to allocate the money evenly across the three years, rather than ramp up costs in the third year – which is difficult for an infrastructure project in which more money is usually used in the final stages [42]. Some cities complain that the larger metros are getting a greater percentage of available funds, while others criticize the smaller cities like Rustenburg for building a BRT, suggesting that the city is too small to sustain the system. There is also a sense that cities are tempted by the availability of the PTISG and are therefore not carefully considering if BRT is the right solution locally. These factors all speak to varying degrees of competition.

One of the more controversial externalities was South Africa's hosting of the 2010 Football World Cup, which took place in July and August in 10 stadiums across nine cites, many of whom were currently in various stages of BRT approval. The 2010 World Cup offered a number of noteworthy opportunities to reform public transportation, formalize the taxi industry and densify the city (see the "State Intervention in Transportation" section). Mega-events like the World Cup act as a "fire in the background", as one South African understood it [39], without which the various development projects would never have been completed, or at least taken longer to finish. South Africa's hosting of the 2010 Football World Cup certainly advanced announced BRT projects, rationalized another transportation advocate, "forcing it to happen" by shortening the project's timeframe and "focusing people's attention in a way that enabled the project's completion". In order to meet the impending deadline of the World Cup, Rea Vaya "was done in such a rush that we didn't have time to talk to anybody", rationalized Colleen McCaul, Project Manager for Rea Vaya Phase 1A between 2008 and 2011. "There wasn't enough opportunity for cities to collaborate because of competition", adds Eddie Chinnappen of Cape Town's MyCiTi. "The mechanisms for learning have not been thought through seriously. But I am strongly of the opinion that cities need to work together and if they do then they won't repeat the mistakes of their predecessors" [89].

One such forum for intercity collaboration is the South African Cities Network (SACN), an NGO established in 2002 to foster communication and cooperation by circulating information, experience and best practice on planning and governance issues between the nine largest South African cities (Buffalo City, Cape Town, Ekurhuleni, eThekwini, Johannesburg, Mangaung, Msunduzi, Nelson Mandela Bay and Tshwane) and provincial and central government. SACN offers cities the opportunity to share their experiences in a somewhat depoliticized atmosphere and provides opportunities for cities to exchange innovation. In September 2013, SACN re-launched the Public Transport Reference Group to provide a peer-learning forum that encourages sharing of good practice and a conduit through which to channel research into policy and practice. Attendance at this first session of the Reference Group was poor with municipal managers largely ignoring this opportunity to meet. Staff at SACN asked the cities if they would still like to have the Public Transport Reference Group and they promised to send representatives to the next meeting. A second session of the Reference Group was scheduled for February 2014. This second gathering was better attended but plagued by political disputes between national and municipal actors, as well as among cities at different stages of implementation.

Officials at the National Department of Transport were interested in creating a platform for the management and retention of technical knowledge and expertise among South African cities. This was a focus in the special BRT workshop at the 2012 Southern African Transport Conference (SATC) in Tshwane, with politicians and technical staff from several South African cities calling for systems to enable secondary adopters to learn from Cape Town and Johannesburg. At the SATC, delegates discussed establishing a repository to manage the varied experiences and knowledge developed during the course of BRT implementation. The National Department of Transport offered to host such an institution. In my interview with Andre Frieslaar, Director of BRT Planning, Traffic Engineering and Transport Planning at HHO Africa in Cape Town, he also told me about the suggestion to develop an archive of experiences. He told me about a meeting he had with Ekurhuleni in June 2012 because "they asked somebody to come and talk about curbside versus median busways and they wanted somebody with experience of the technical components of BRT". Ekurhuleni had visited Johannesburg and were planning to visit Cape Town, but Frieslaar explained that officials found these technical conversations more useful than their earlier political meetings [86].

Perhaps through the shared practice of learning, BRT actually brought rival cities into conversation with one another. There is evidence of more substantive exchange beginning to take place between the early adopters, Cape Town and Johannesburg, and the later adopters such as Rustenburg and Tshwane. Eager to avoid reproducing the system errors of its predecessors, in 2011, rather than visiting Bogotá, Rustenburg travelled to Cape Town and Johannesburg to meet with officials and ride the MyCiTi system. One of the lessons Rustenburg took from operational systems in South Africa is to "take a broad and integrated approach from the beginning" and focus on the variations and principles of BRT, rather than "being engineer-led" [75]. Political and technical officials in Rustenburg are also trying to contemplate how to avoid the difficulties with the taxi industry that derailed the BRT in Nelson Mandela Bay. The affected taxi operators are particularly eager to duplicate the negotiated outcomes in Johannesburg, which promised affected taxi operators profits regardless of ridership figures. Operators demonstrate this association by frequently, perhaps accidentally, dubbing the Rustenburg system "Rea Vaya" rather than BRT. Along the same vein,

Tshwane utilizes footage of the buses and stations of Rea Vaya in its promotional videos. These anecdotes are indicative of the possibility for more substantive exchanges across South African cities as BRT implementation continues.

The Municipal Context of BRT Circulation

This previous section focused on the national South African context and influences guiding BRT adoption. Within South Africans cities, however, reaction to BRT differed considerably. Figure 5.2 depicts the process through which the six South African cities of the study adopted BRT policies. These stages are organized according to three criteria: the knowledge stage, in which cities pursue the details of the BRT concept as determined by the act of undertaking a study tour; the decision stage, in which cities accepted the BRT concept as measured by the act of council approving BRT plans; and the implementation stage, when BRT materialized in the local city or as assessed through the act of construction. While these stages are generally assumed to be linear, this figure reveals the variations in policy circulation, which leads to differences in adoption and implementation. For example, Johannesburg underwent the knowledge and decision stage simultaneously, followed immediately by implementation. Both Tshwane and Nelson Mandela Bay followed the same process, but project progress halted in the implementation stage. Meanwhile, Cape Town experienced the knowledge stage and then waited a year before undergoing the decision and implementation stage

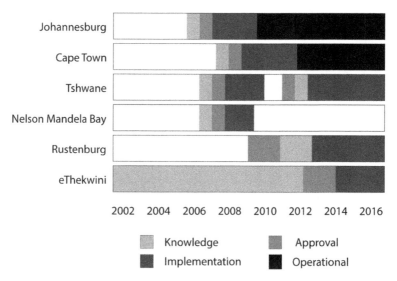

Figure 5.2 Learning process across South African cities.

concurrently and was successful in adopting BRT. Rustenburg made the decision to adopt BRT before undergoing the knowledge stage and then did so concurrent with the implementation stage. Interestingly, eThekwini waited 10 years between the knowledge stage and then underwent the decision and implementation stage simultaneously.

One explanation as to why BRT adoption varied across South African cities is because of local leadership, which in some cities endorsed adoption and in other instances, stalled it; and in still others, was entirely absent letting the opportunity slip past. This institutional context supporting BRT adoption was not just of a top-down persuasion, there was similar pressure from within city government. This pressure was often described in interviews as the critical role of a political champion in advocating for BRT. In reaction to the national legislation devolving public transportation to cities when they are ready (see the "State Intervention in Transportation" section), Johannesburg, for example created a specialized department for transportation, distinguishing it from the planning unit to elevate the importance of improved mobility. A position for Moosajee, who was so instrumental in advancing BRT strategies, was part of this restructuring. Moosajee explained how beneficial these relationships were in driving the project forward. She realizes that "the actual dynamics in cities are different because of different leadership and different ways of doing things" and "no two cities are the same" so "each city has to work out its own plan for how to build BRT" [22]. This created a local context for policy adoption, such that cities like Johannesburg with Moosajee and Cape Town, in 2006 led by Mayor Helen Zille, and since 2012 led by MMC Brett Herron, are more likely to adopt circulated policies than Tshwane, eThekwini and Nelson Mandela Bay, that do not have a strong political champion.

Figure 5.2 indicates that eThekwini was initially reluctant to implement BRT and in spite of having visited Bogotá, officials remain skeptical as to whether it could be viable locally. Notwithstanding these reasonable apprehensions, in 2012 eThekwini Council approved plans to proceed with three lines of a BRT system. This lag can be interpreted as inner-city political and governmental limitations because despite eThekwini's propensity for experimentation, there were no local political champions to advocate for transportation improvements. The city council is managed through sector committees and the transport committee, which is clustered with economic development and planning, is particularly chaotic. The committee has about 40 members without an individual politician to manage the group or champion its goals. Thami Manyathi, Head of the eThekwini Transport Authority since 2010, calls this the city's "Achilles Heel", explaining, "We don't get a quick turnaround on some of our issues as would be the case if we had a dedicated committee and a strong political champion" [49]. I asked Manyathi why eThekwini had finally decided to adopt BRT? He responded that the opportunity to build a BRT is "too big a temptation for cities to resist" and because of the experiences in Johannesburg and Cape Town, he thinks eThekwini can build a viable, sustainable BRT system. Another representative from the eThekwini Transport Authority further clarified, "the political dimension where the politicians are asking why we are so slow to build BRT" is perhaps too persuasive to ignore. Politicians are telling their technical staff, "Cape Town and Johannesburg have a BRT, when are we getting a BRT? [47]".

This lack of consistent leadership also limited Tshwane's advancement of BRT policies. Tshwane had three mayoral members between 2008 and 2012, which according to Hilton Vorster, Director of Traffic Engineering and Operations from 1975 to 2013 and Project Manager of BRT until its stoppage in 2010, explains, "makes it very difficult because you don't have continuity". Tshwane's Department of Transport has also been challenged by instability. Between 2010 and 2012, Tshwane has had four heads of department – two permanent and two acting directors. Vorster explains how he convinced four different directors of the attributes of BRT, but as soon as he got approval from one director, they were replaced and the cycle repeated. "Continuity has been a major challenge for this project" and without "political buy-in" and "constant top management support", Tshwane's BRT "never got off the ground" [42]. Since policy mobilities is an inherently social process, such inconsistency in leadership limits the decision-making process regarding how to act with the information.

These political dynamics introduce a new challenge regarding the management of the transportation function in South African municipalities. A number of South Africans mentioned the need to establish a municipal transportation authority to manage the construction and operations of an integrated transportation network. Under the auspices of a transportation authority, "the city moves from being a glorified roads department to being a proper mass transport authority like in Singapore or Hong Kong or New York City", recounted Ibrahim Seedat, Director of Public Transport Strategy at the National Department [38]. In October 2012, Cape Town became the first municipality to take over responsibility for rail in accordance with the National Land Transport Act. In February 2012, Metrorail, South Africa's commuter rail network, characterized by a lack of capacity, outdated rolling stock and reactive rather than preventative maintenance, announced plans to transfer control over the rail services in Cape Town to the newly formed transportation authority [55]. There was substantive concern, however, that these authorities would add another layer of governance, inhibiting the establishment of robust municipal transportation departments. Philip van Ryneveld, a staunch advocate of decentralization, argues, "You already have a city government that is the transportation authority, so you don't want to create another authority outside of that because then you are just fragmenting responsibility" [54]. In Gauteng, there was apprehension that a transportation authority would be managed by the province, thereby challenging the municipal centrality sought through BRT. The introduction of BRT was hastened by this political move towards decentralization, and likewise its introduction furthered municipal dominance in the transportation function.

These cases depict the political interactions between officials within the same city and the impact of local politics on the adoption of BRT. While all South African cities faced similar opportunities and limitations, the variations in local politics influenced each of their implementation of BRT.

Conclusion

This chapter sets out to explore the international, national and local landscape influencing BRT adoption in South African cities. I examined the wider political relations, which include the aspiration to strengthen ties with another political system and competitive tendencies between cities to be seen as the most pioneering and capable municipality, as well as the intra-urban processes driving the adoption of peripatetic policies. I have reinterpreted the circulation and adoption process as a multifaceted and irregular progression, in which some cities accept BRT, others reject it and some take longer to make their decision. Certainly, no policy is universally accepted or rejected. Rather, policy adoption practices are rooted in these institutional frameworks and power structures and shaped by local conjunctures and contextual realities.

In addition to detailing the happenings within South African localities, this chapter has examined the way in which these exchanges develop and modify relations between policy exporters and importers, as well as among importers. In some instances, policy mobilities improves the relations between cities by creating greater opportunities for collaboration, connecting localities through knowledge exchanges, apportioning financial resources and staff capacity; or cities might see the development of international prestige as of benefit to all localities, presuming that it will be easier for other municipalities to follow a local case of borrowing. Many times, however, innovation drives competition, with cities battling over inadequate capital and labor, national politics and global prestige produced by local dominance as pioneer and innovator; cities might compete for trade and investment opportunities and associated donor interest and sponsorship; or perhaps city branding and marketing strategies drive regional antagonism. In other instances, interurban competition can be advantageous – it can inspire cities to build experimental BRT systems, which better serve the needs of riders and become best practice around the globe.

South African cities were presented with the BRT model and each locality made a choice regarding the implementation of BRT. Variation in uptake differs because of the local political context, which includes a range of structural and non-structural issues including economic competitiveness, geographical size and financial capacity, as well as the political perspicacity of particular policy actors and associations in promoting circulated policies. The decision to adopt mobile policies is linked to the unequal power relations between places and the uneven distribution of financial and technical opportunities to act on circulated knowledge. This explains why the implementation of MyCiTi in Cape Town took on a more technical nature, while in Johannesburg, passionate politicians like Rehana Moosajee ensured that BRT remained committed to social transformation. These arguments illustrate that policy mobilities is more than international policy consultants promoting international policy models, but is rather part of the complex processes of intergovernmental mobility and competition. These global spatialities of local policy adoption are also profoundly influenced by temporality, and as I will argue in the next chapter, policies circulate through multiple temporalities, which also explains the shifting sociopolitical landscape of the South African city.

Chapter Six
Repetitive Processes of BRT Adoption

Introduction

Since 2006, BRT systems have opened in Johannesburg, Cape Town, Tshwane, and Ekurhuleni with four other cities – eThekwini, George, Nelson Mandela Bay and Rustenburg – in various stages of planning and implementation. While this rapid and complete transformation of the South African urban transportation network may appear fast, several local agents pointing to its speedy implementation as one of the most attractive features of BRT, further investigation reveals a lengthy and protracted process riddled with experimentation and failure. This chapter investigates the South African city's transportation history to consider how prior experiences with similar types of urban transportation – trams, trolleybus and exclusive busways – inform ongoing decisions to adopt BRT. When the Bogotá model of BRT arrived in 2006, "it wasn't the first time that South Africans had heard of BRT", recalls Pauline Froschauer, Director at Namela Consultants and project manager for Rustenburg Rapid Transport, rather "it was just one of those things that never really stuck" [75]. In departing from the prevailing logic in the policy mobilities literature, which understands the policy world to be comprised of "fast policies", these arguments display a learning process that is lengthy and drawn out, incremental and at times delayed.

This account of multiple temporalities builds upon the burgeoning literature on policy mobilities, which interprets the fast institutional re-engineering of off-the-shelf prefabricated best practice policies to shorten policymaking cycles. The case of BRT adoption in South Africa offers an occasion to consider the

How Cities Learn: Tracing Bus Rapid Transit in South Africa, First Edition. Astrid Wood.
© 2022 Royal Geographical Society (with the Institute of British Geographers). Published 2022 by John Wiley & Sons Ltd.

process of policy mobilities as constant, gradual, creeping, at times sticky, and other times loitering, not something prompt and hurried. This chapter advocates a reappraisal of the mobilities process to survey the pertinent earlier experiences advancing local development. By tracing the chain of BRT transmission from its initial forays through its subsequent failures in South Africa, this research adds a chronological approach to the implementation of BRT by shedding light on the process of transnational policy flows. Slow policy dissemination has been understudied in previous research, mostly because it is difficult to locate evidence of incomplete mobility. The example of BRT adoption across South Africa supports my claim that policy exchanges are far longer in duration than usually assumed, and that interactions from bygone years can be fundamental in shaping ongoing mobilities. The learning process, as interpreted here, is prolonged and sequential with a number of encounters necessary before it takes root. Rather than simply a process of rapid reproduction, consideration for the history of policy ideas reveals that policy mobilities is in fact stubbornly dependent on the local context.

This chapter departs from the theoretical discussions of the circulation process as a contemporary phenomenon, by illustrating the long history of BRT awareness and the role of previously unsuccessful instances of circulation to inform and prepare South African cities for the most recent iteration of the model. The next section places these research objectives within a wider discussion outlining South Africa's long history with transportation innovations including trams and trolleybuses. The main empirical section then situates three previous encounters with BRT – the first, a 1973 conference report, the second, as part of the post-apartheid obsession with Curitiba and the third, the enthusiastic promotion of Bogotá's Transmilenio – within a failed attempt in 2002 to build a BRT lane along Klipfontein Road in Cape Town. These seeds were fundamental in fashioning a fertile ground for future championing of BRT. The final argument looks at three other transportation innovations flowing through South African cities – personal rapid transit (PRT), the monorail model and Gautrain – two of which were considered but then ultimately rejected – and provided useful experiences for the introduction of BRT. These examples of policy failure demonstrate the way in which even unsuccessful transfer is critical in shaping uptake, and provide further evidence of the multiple temporalities of policy mobilities. In concluding, the genealogy of BRT circulation presented here makes the case for the importance of appreciating historical policy mobilities and the multiple temporalities through which policies flow.

Tracing Transportation Innovation in South Africa

South African transportation history is composed of instances of adopting the latest transportation technology, before its sudden and complete replacement by the next best practice. Architects and urbanists frequently pursued antidotes to local urban challenges in British town planning models, which diffused through the global hegemony of western imperialism and were put into practice with relative ease (King 1976; Lemon 1991). The nascent mining outposts and port villages of South Africa followed the example of their European counterparts, by separating the worker from their place of work and the factories from the inner city, all of which necessitated the development of a viable means of public transportation. In the 1920s, the British garden city model, which called for efficient, decentralized, residential areas to replace the dirty and crowded cities of industrialized England, was replicated as Pinelands, Cape Town (Wood 2019a) and later inspired the development of segregated townships in Langa, Cape Town; Lamontville, Durban; and McNamee, Port Elizabeth (Maylam 1995). At times, governance drove technological transformation, and in other instances, profit facilitated the localization of best practice. It is within

Figure 6.1 Horse-drawn tram in Johannesburg.

In April 1889, the South African government granted Sigismund Neumann a concession to construct and operate Johannesburg's first tramways. He laid the tracks in the public streets of Johannesburg in front of this then government building. Johannesburg's first tram began operating on 2 February 1891 between Height's Hotel, Commissioner Street and Jeppestown. The service was so successful that after 9 June, it was extended to other areas of the city (Norwich 1986: 24).

this context of mobility that we can demonstrate how new and recycled ideas relating to transportation were introduced, relocated, replicated, and replaced across South African cities.

In Johannesburg, the tale of transportation begins with horse-drawn streetcars in April 1889, when the government granted the Johannesburg City and Suburban Tramways Company a concession to construct and operate the city's first tramways. Services began operating in February 1891 between Jeppestown and Fordsburg to transport the White working-class miners residing there (see Figure 6.1) (Norwich 1986; Spit and Patton 1976; Van Onselen 2001). Similar services opened in Cape Town in 1863 (Gill 1961), Durban in 1880 (Jackson 2003), Port Elizabeth in 1881 (Harradine 1997; Shields 1979), and Pretoria in 1897 (Joyce 1981). The sensation of the streetcar was short-lived as the maintenance of the horses proved difficult.

Electric trams soon replaced their horse-drawn counterparts, with Cape Town being the first to introduce a service between Adderley Street and Mowbray Hill in 1896 (Joyce 1981). Their success in Cape Town led to their installation across South African cities: In 1897, Port Elizabeth opened its electric tram; Durban opened in 1902 moving people from the city center up Florida Road to the Berea (Jackson 2003); Johannesburg electrified its horse-drawn service in February 1906 (see Figure 6.2) (Sey 2012; Spit and Patton 1976); and Pretoria electrified its tram in 1910 (Joyce 1981). For the most part, private

Figure 6.2 Electric trams in Johannesburg.
This open-top tram, moving east at the corner of Loveday and Market Street, demonstrates that passengers were either exposed to the elements atop or trapped in a sardine-tin below (Norwich 1986: 38).

entrepreneurs built these services and once profitable, the city acquired the system. Electric tram services were the global model of excellence around the world and therefore the presumed remedy for the overcrowding and congestion in South African cities.

The triumph of the electric tram was also fleeting. By the 1920s, tram patronage declined significantly as private car ownership increased and diesel buses captivated the sprawling suburban market (Rosen 1962). The once beloved trams were now noisy, uncomfortable and slow (Joyce 1981); whereas once they facilitated the rapid expansion of the city, now they were impeding its growth. The Spencer Commission, sponsored by Johannesburg to investigate the long-term viability of the tram network, reported that the density per acre was substantially lower in the city's northern suburbs than British cities of comparable size, and as a result, public transportation links were unlikely to be longstanding (Beavon 2004). Their findings led to the subsequent closing of all tram services across the country, with the same eagerness with which they were first constructed. Both Cape Town and Pretoria terminated services in 1939 (Gill 1961), Port Elizabeth in 1948 (Patton 2002), Durban in 1949 (Jackson 2003), and Johannesburg; the last city to lay tracks in 1948, was also the final city to terminate services in 1961 (Hart 1984).

These patterns repeat for trolleybuses, dubbed the "trackless tram", in South African cities during the 1930s and then similarly replaced – in Cape Town in 1964, Durban in 1968 and Johannesburg in 1986 – by diesel buses. Just like the tramcars, the trolleybuses were also imported from Europe, with those in Johannesburg originally made by the Italian firm Alfa Romeo (Joyce 1981). In the 1980s, trams briefly re-emerged as the transportation solution of the future with calls to reinstall municipal tram networks (*The Economist* 1989). Diesel buses, however, dominated the urban landscape until the expansion of the minibus taxi industry, which when deregulated in 1986 demonstrated the failings of buses to service the extraordinary demand and low-density urban form.

Ron Haiden, BRT Manager of Infrastructure and Development in Cape Town, interprets these various cycles as part of a long history of transportation recycling, in which local South Africans introduce the latest transportation innovation only to replace it when a more modern solution arrives. "They look different and we give them different names. Light rapid transit is the old tramway system with new technology", Haiden explains. "So, how does learning happen? It is through evolving ideas. Understanding those ideas and then interpreting them for application in a different socioeconomic, climatic condition" [55]. The introduction and removal of each of these services, reflects the socio-spatial context in which transportation was provided, and speaks to the wider scope of this chapter: how does South African experimentation with transport innovation inform ongoing BRT adoption? And, more theoretically, how does previous experience with a similar innovation inform local adoption practices? These considerations will be used to analyze the multiple temporalities through which BRT arrived in South African cities.

Planting the Seeds of BRT in South Africa

Interviews reveal that by 2006 a substantial portion of South African policy actors were aware of BRT with several working on a variety of alternative urban transportation interventions shaped by BRT encounters. In a number of instances, interview respondents had seen BRT projects on study tours to Curitiba, and in other cases, they learned of it through direct involvement in the failed attempt in 2002 to build a BRT route along the Klipfontein Corridor in Cape Town. Still others learned of it at overseas universities. The results reveal that 66 percent admitted to knowing about BRT in 2006, while just 34 percent learned of it following the events of 2006. Figure 6.3 indicates that responses were dispersed across political and technical sectors, as well as evidenced throughout their structural hierarchies. In spite of this knowledge, there were no successful BRT projects in South Africa prior to Johannesburg's Rea Vaya opening in 2009.

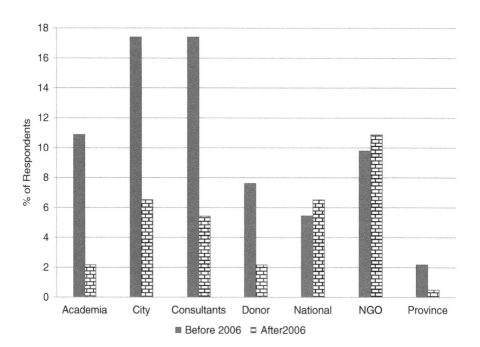

Figure 6.3 Knowledge of BRT adoption in South Africa.

Previous experience with public transportation led some South Africans to argue that BRT implementation has been slow, since many of them had already spent years planning for a major urban transportation system. Several had seen BRT projects on government sponsored study tours to Curitiba. Others had even tried to initiate BRT projects locally. Transportation experts frequently mentioned their familiarity with BRT prior to 2006. Ron Haiden explains, "I think people writing the history books will look to Curitiba as the beginning of BRT and I think that is widely acknowledged. But where do ideas come from? There is nothing new under the sun. It is just different technologies" [55]. Haiden is referring to several innovative transportation projects – a reversible lane on the R27 to Table View, the Bus-Minibus Taxi Lanes on the N2 Highway in Cape Town. He also planned to apply a version of BRT along Symphony Way in Blue Downs – similar to the Bogotá model of BRT – which were designed and operationalized in Cape Town but never implemented, because of a lack of political and financial support [63]. Haiden argues that this is because South Africans "didn't write about it and we didn't shout about it" [55]. His arguments were corroborated by a number of other South African policy actors, who describe BRT as the current fashion but attribute its origins to earlier flows [5; 11; 36; 38; 68; 86]. Rehana Moosajee concludes, "Everything is an international idea at the end of the day. There is nothing that is totally original" [22]. These respondents refer to the long history of South African policy actors visiting and learning from Curitiba's land-use and transportation innovations.

Some policy actors were concerned that BRT approval was hasty, arguments which are suggestive of a "pragmatic copying" of best practice quick fixes often characterized as "fast policy" (Peck 2002). These interview respondents reference the Public Transport Infrastructure and Systems Grant from National Treasury, the 2010 Football World Cup, and legislation devolving responsibility for bus, rail and taxi to cities as catalysts enabling BRT promotion (see the "State Intervention in Transportation" section). "It happened so fast", explained Sharon Lewis, Executive Manager of Planning and Strategy for the Johannesburg Development Agency, the organization responsible for managing the construction of Johannesburg's Rea Vaya [2]. This atmosphere was also expressed by Pauline Froschauer, leader of the Rustenburg BRT team, who simplified the hasty nature of the project as "the sooner, the quicker, the faster, the better" [75]. Within this expedited timeframe, decisions were made in reference to historical and local examples of urban public transportation. While this hastened the project along, some of the subsequent problems with BRT operations are symptomatic of that speed.

For the most part though, local policy actors described the project as slow, since many of these actors had already spent years planning and laying the groundwork for a major transportation intervention like BRT [8; 46]. In my interview with Colleen McCaul of Johannesburg's Rea Vaya, she argues against the assumption that BRT was "fast" recognizing however that "it could not have happened without there being so much fertile ground in the city" [5] (see also, McCaul 2008).

Lloyd Wright, a world-renowned expert in BRT, agrees with McCaul describing the process as slow – "Once you plant a seed, it's tough to get that seed to grow but very often you see replication very quickly. Today in Colombia, ever since Transmilenio was implemented in 2000, we have implemented eight projects. That is fairly quick in terms of public transport infrastructure" [27]. Respondents assert that this "fertile ground...prepared people at the political level to accept BRT" [64]. Several interview respondents substantiated this concept of a fertile ground, arguing that ideas of density and transit-oriented development have been flowing through the South African urban terrain for some time.

A number of South Africans base their seemingly speedy support for Bogotá-style BRT within concepts of density, growth management and transit-oriented development, which they wrote about in policy documents and reports years prior, and later used to inform the introduction of BRT in South Africa (e.g. City of Cape Town 2003; HHO Africa 2000, 2004). Several attributed their knowledge of BRT to experiences at university or employment opportunities in Europe and the United States, where they encountered these concepts and since the 1980s have been re-introducing them in South Africa [8; 61; 69; 87]. John Jones, Director of Engineering at HHO Africa explains, "Transit-oriented development goes back here for me to the 1980s, when I was at Berkeley doing my graduate work, and Andre was in Texas in the 1990s. That was new then and in 2003, it was new here. It became the flavor of the month but it was donkey's years old" [87]. For some, because BRT had come through previously and then returned, they inferred that it must be the best solution [36]. When BRT re-emerged in 2006, "the fertile ground then had some fertilizer strewn on it" [11].

Following their admission that BRT was in fact known long before 2006, I asked interview respondents, "Is this the first instance when something like BRT has been introduced from elsewhere and implemented?" Most respondents readily admitted that this was not the first time a transportation innovation had moved through the urban terrain, nor was it the first instance of BRT arriving in South Africa. Policy actors told me about an attempt in 2002 to develop BRT in South Africa along the Klipfontein Corridor in Cape Town, but for financial, political and technical reasons, the project never materialized [38; 68; 70; 89]. Johannesburg-based respondents presented the example of the monorail in 2007, when a Malaysian consortium offered to build, finance and operate a R12 billion monorail from Johannesburg center to Soweto [8; 17; 22; 38; 39]. This led several policy actors to conclude that BRT was just another "policy fashion" in which wealthy countries disperse "new tools" through the global south [55].

BRT thus has a longer, more convoluted history in South Africa than often assumed by local implementers. The remainder of this chapter analyzes the importance of previous experiences with versions of sustainable transit in advancing BRT discussions. It interrogates the role of multiple temporalities of learning – gradual, repetitive and delayed experiences with notions of BRT – in influencing ongoing support.

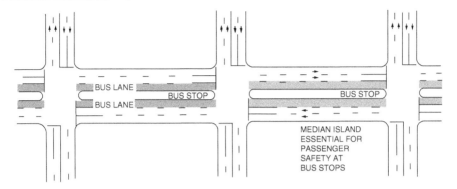

Figure 6.4 Exclusive curb lane on city streets.
This is an image from Louis de Waal's report, Bus Lanes and Bus Priorities (1973: 7) presented to delegates at the South African Institution of Civil Engineers quinquennial conference in Johannesburg. Notice the way in which he directs the buses through the center of the roadway, a form later replicated in BRT systems around the world.

Gradual Processes of Learning

The earliest written discussion of BRT in South Africa is found in a 1973 conference paper, "Bus Lanes and Bus Priorities", proposed by Louis de Waal, a transportation engineer working in Cape Town and presented to delegates at the South African Institution of Civil Engineers Quinquennial Conference in Johannesburg (de Waal 1973). In the paper, de Waal offered an at-grade solution to South Africa's increasing car ownership and spatial fragmentation – a bus running through the median of the roadway moving with the fluidity of a train – and he ascribed its potential success in South African cities to its achievements elsewhere (see Figure 6.4). In particular, he references bus lanes, bus priorities, comfortable buses, and carefully sited bus stops in Chicago, Los Angeles and New York in the United States, as well as London, Nottingham and Southampton in the United Kingdom. Within South Africa, he directed attention towards the exclusive bus lanes in Johannesburg. These lanes were delineated from ordinary automobile traffic by solid white lines, white plastic posts and letters painted in the center stating "bus only", concepts currently used in Cape Town's MyCiTi and Johannesburg's Rea Vaya.

The first evidence of a BRT-like intervention is in central Johannesburg, where a special bus-only lane was established to increase the flow of public transportation during the peak hours. At another intersection, priority was given to buses turning right because it was found that traffic formed during the morning commute. Rea Vaya currently operates in both of these areas. References are made to these schemes in the records from the 1973 conference of civil engineers, suggesting that even then South African cities were learning alongside one another. Written comments also associate these South African interventions with a new service launched in London – a network of limited-stop bus services utilizing a reserved lane called "speedbus", linking those suburban areas not served by the underground system. In the conversation that followed, the conference chairperson mentioned that exclusive bus lanes were experimented with on the freeways in Cape Town, but were considered by the local bus company to have failed because of insufficient ridership. The lanes described closely resemble the Bus-Minibus Taxi Lanes on the N2 Highway put into practice in Cape Town more than a decade prior to the 2006 BRT discussions.

Many of these early initiatives were developed by South African engineers and planners, aware of best practices such as BRT but also attentive to governmental limitations with regards to construction, management and operations, which made it difficult to fund major capital improvements including formal transportation systems. Andre Frieslaar, Director of BRT Planning, Traffic Engineering and Transport Planning at HHO Africa in Cape Town, recalls "that famous email from Louis de Waal" sent in the late 1990s describing the attributes of BRT and he thought, "Could this be our next major opportunity?" This was long before the attempt to build BRT along Klipfontein Corridor or the campaign for Rea Vaya. From that email, "I got my hands on a presentation that Lloyd did and I learned about BRT from that", recalls Frieslaar, "Then I had Lloyd's details from that email and so I contacted him to ask him some questions so that is how we first met" [86].

Several South Africans had also learned of and experienced innovative transportation solutions through educational or employment opportunities in Europe and North America. John Jones remembers not only reading about but also experiencing transit-oriented development while at the University of California, Berkeley in the 1980s [87]. Likewise, in Johannesburg, Bob Stanway recollected his experiences working on the Jubilee Line of the London Underground when developing Rea Vaya [8]. These contentions imply a learning process that is gradual and subtle, even sluggish, in which actors slowly warm to the idea. Several of these South Africans remained committed to the principles of good public transportation and called upon these earlier encounters when introducing BRT in 2006.

Repetitive Processes of Circulation

The stream of international urban transportation solutions increased after apartheid ended in 1994; ideas about efficient cities and moving people rather than vehicles fit well within post-apartheid transformation, which called for a reconsideration of South African urban form. There was a lot of "policy thinking" during this period because of the unbanning of the ANC, forthcoming elections and the composition of the new national constitution, illuminated Colleen Mc-Caul in Johannesburg. She described the South African urban context as "a very fertile sort of place for discussing new ideas, and people were searching for ways to deal with South African city problems like taking very poor people very long distances" [5].

During this time, Curitiba became the model for innovative land-use models, workfare programs and urban transportation. Although concepts of BRT, such as buses running in exclusive trunk median bus lanes and stations with controlled entry/exit points can be traced back to 1937, when Chicago outlined plans for express bus corridors, Curitiba's Rede Integrada de Transporte, which opened in 1974, is reported to be the first complete BRT system. Many of those currently involved with BRT in South Africa trace their learning to Curitiba and the study tours through which ideas and practices found their way to forefront of discussions in civil society. Professor Philip Harrison, South African Research Chair in Spatial Analysis and City Planning at the University of the Witwatersrand in Johannesburg, describes the example of Curitiba as "iconic here in South Africa", [4] and Colleen McCaul calls Curitiba "the mecca of international planning ideas long before Bogotá" [5]. Curitiba remained a common reference point as evident with several South African planners and politicians repeating Curitiba's slogan, "Think Rail, Do Bus" in reference to ongoing BRT support [38; 40].

Ideas frequently entered South Africa from Curitiba through local talks and seminars [5; 64; 88]. One such instance was in 1999 when Jonas Rabinovitch, an architect and urban planner with the Curitiba Research and Urban Planning Institute (IPPUC) and author of the World Bank Case Study on Curitiba, presented Curitiba's success with BRT to Parliament's Transport Committee [23; 62]. Rabinovitch clarifies in his case study, "We do not intend to depict Curitiba's approach as a turn-key model for reproduction elsewhere. Cities are too different in context and resources for such wholesale adaptation" (Rabinovitch and Hoehn 1995: 9). Instead, his objective was to help South African cities understand how Curitiba developed its innovative system and think about what is possible locally. Madeleine Costanza, who worked for the International Institute for Energy Conservation (IIEC) from 1998 to 2002 was responsible for bringing Rabinovitch to South Africa. "The BRT aspect was only one portion of the message", she adds, "and even then I am not sure I clearly understood how instrumental it could be for a city in a developing country" [23].

For the most part, learning from Curitiba took place through a plethora of study tours. Professor Roger Behrens, Director of the Centre for Transport Studies at University of Cape Town (ACET), describes South Africans' fascination with Curitiba's best practices through the 1990s as a "love affair". "We had a large number of delegations travel to Curitiba in the mid-1990s, not so much to do with transport per se but more the integrated land-use transport success of Curitiba" [11]. One such study tour to Curitiba was instrumental in helping South Africa prepare the 1996 White Paper on Transport (National Department of Transport 1996). Professor Jackie Walters, Chairperson of Department of Transport and Supply Chain Management at the University of Johannesburg, organized a tour for himself and those composing the White Paper to visit Curitiba. He remembers thinking that BRT would be an ideal solution in the South African context, because it was cheaper and more flexible than light rail and thus could easily be adapted to the changing shape of the post-apartheid city. These experiences were instrumental in the composition of the White Paper, which provided a framework for subsequent transportation policy in South Africa [74]. Other outcomes from this learning were evident in various planning reports including for example the Koeberg Road Management Strategy, a traffic and transportation report published for the City of Cape Town in 2000 by the same transportation engineering firm that sponsored the 1973 report mentioned earlier (HHO Africa 2000, see also, 2003, 2004, 2007b).

These repetitive exchanges were necessary in creating a fertile ground for the ongoing application of the Bogotá model of BRT, by laying seeds of the innovation within the South African context. Wright spoke of the need to present the achievements of BRT a number of times in South Africa. "I remember presenting Curitiba often in the mid-1990s and it was interesting, but the problem is that when you only have one city then people just dismiss it and say that is just a one-off. But when there are ten or twenty of these projects, then you start building up a critical mass and everyone wants to do something" [27]. This suggests that policy learning is recurrent and takes place through a number of temporal episodes, South Africans became familiar with the possibilities for BRT. Although BRT "didn't happen right away", as Paul Steely White from Transport Alternatives put it, these various experiences inspired future thinking about urban development and transportation planning [41].

Delayed Processes of Adoption

In 2000, Bogotá surpassed Curitiba as "the international exemplar for developing world countries", recalls Roger Behrens, Director of ACET in Cape Town [11] (see the 'Forming the Bogotá Model of BRT' section). Bogotá's Transmilenio was the "game-changer", because it was the first comprehensive BRT system built in a major metropolis to replace an unregulated and unruly transit industry, and thus considered a more analogous example to South Africa than Curitiba. In addition to providing an example of the infrastructure of BRT – dedicated median lanes, iconic stations with high-floor platforms, off-board fare collection – Transmilenio importantly also demonstrated methods for incorporating existing transit operators, a major issue for urban transportation in South Africa (see the "BRT and Taxi Transformation" section). That was when "we saw the second love affair happen", which included a number of visits by international transit experts with experience building BRT and a host of government-sponsored fact-finding visits to Bogotá [11] (see Figure 3.1).

By this time, South African cities were already in the process of improving their urban transportation systems: Johannesburg was realizing the Strategic Public Transport Network (SPTN), a plan first approved in 2003, aimed at introducing a more coherent, organized, grid-like shape to the existing public transportation network. The SPTN, which included 250 km of public transportation priorities on existing lanes, was designed to enhance the speed and service of existing transportation systems (de Beer 2004a, 2004b, 2005). Cape Town was also introducing a suite of public transportation interventions, including a reversible lane on the R27 to Table View and the Bus-Minibus Taxi Lanes on the N2 Highway. As early as 1999, the city had plans to pioneer a version of BRT along Symphony Way in Blue Downs, similar to the Bogotá model of BRT but based on their learning from Grenoble, France. None of these interventions were as radical as BRT; not one called for the formalization of the taxi industry, nor did they address the role of the municipality in system operations. While local planners and politicians were aware of better solutions elsewhere, planning is a lengthy, laborious process that relies on a host of local financial, institutional and political conditions. Failure is also a critical component of delayed endorsement.

In 2002, vague discussions regarding introducing a new public transportation service along Klipfontein Road in Cape Town were solidified at the then newly established Centre for Transport Studies at the University of Cape Town (ACET). ACET hosted an open lecture series on public transportation in which global experts were invited to contribute to ongoing discussions. In September 2002, Lloyd Wright, then Regional Director at ITDP in South America, presented the recent developments of BRT and non-motorized transportation systems with a special focus on Bogotá's Transmilenio (Wright 2002). Roger Behrens remembers because this was the first in the lecture series. "He was passing through and an American by the name of Madeleine Costanza contacted us for him to give an open lecture, which we did in 2002, and sitting in that lecture was an advisor to the then-Minister for Transport called Robbie Robertson" [11]. The Minister was impressed with the achievements of Transmilenio and he shared these thoughts with then Provincial Minister of Transport in the Western Cape, Tasneem Essop. She invited Lloyd Wright to make a similar presentation to her office and she was sufficiently impressed to undertake a study tour to Bogotá, in which she took several industry leaders and city officials [11]. Wright recounts his experiences with Essop, "In the case of Tasneem Essop, it was 'Let's pack up my team. Let's go to Bogotá and let's make a project out of it'. She was sold on it within 15 minutes" [27]. She was very enthusiastic and eager to learn how to bring BRT to South Africa. Plans for Klipfontein were launched at ITDP's "Building a New City Tour" in Cape Town [59].

Klipfontein Road, a double-laned road running through the southeast from Mowbray through Athlone, Gatesville, Gugulethu, Nyanga and ending in Khayelitsha, was slated to be Cape Town's and South Africa's first BRT route. Like Transmilenio, the Klipfontein BRT would run in the median-lane with dedicated stations and prepaid boarding. The plan was to transform 20 km of a mostly desolate stretch of road into an economic heartland through transit-oriented development. The wide grassy strip along the busy Klipfontein Road would be redeveloped for shops and cafes to serve the tens of thousands of BRT users. Plans for Klipfontein were described as a "mobility strategy" (Williams 2003a) for "social engineering" (Williams 2003b) in the overlooked southern suburbs of Cape Town. Essop imagined Klipfontein Corridor becoming a destination stating, "We want to inject life, inject hope into the Klipfontein Corridor...We want to turn it into a social and cultural hub again...Imagine them [township residents] taking an evening stroll to a street cafe on a paved street mall that is well-lit and secure" (Williams 2003b).

The Klipfontein Corridor never materialized for a number of financial, political and technical reasons. The project was exceptionally political and despite years of effort, in May 2004, Essop was reappointed to Provincial Minister of Environmental Affairs and Development Planning, and another individual took up her portfolio and her BRT project fell by the wayside [11; 27]. Technically, the project was a challenge because the two-lane road through Mowbray was particularly narrow and there was insufficient space to reallocate to BRT. The community protested claiming that Cape Town wanted to tear down historic buildings, and while the technical plans could have accommodated the narrow streetscape by allowing the BRT to move in mixed traffic through Mowbray, such technical challenges required political will to overcome [51]. As challenges arose, this seemingly pleasant partnership between Cape Town and the Western Cape Province fizzled out. In 2003, Cape Town was permitted to plan for public transportation initiatives, while the Western Cape Province was responsible for funding the infrastructure. When Cape Town failed to match the financial investment of the Province, problems arose and the relationship suddenly collapsed [68].

But the project did not so much fail as morph into contemporary plans for BRT in Cape Town. These seeds materialized as learning experiences and contacts. When the Klipfontein project fell apart, Lynne Pretorius from Cape Town and her counterpart in Western Cape Province, Zaida Tofie, founded a consulting firm, Pendulum Consultants, specializing in transportation projects. They were appointed in 2008 to do the operational planning for Cape Town's MyCiTi Phase 1. Pretorius and a number of other South Africans involved in BRT, including Maddie Mazaza, Director of Transport in Cape Town, conclude that Klipfontein was instrumental in creating a fertile ground for ongoing BRT projects [38; 49; 61]. Mazaza argues that the starter route simply shifted from Klipfontein to the R27 and Klipfontein became Phase 2 instead of Phase 1 (Mazaza and Petersen 2003).

The experience of Klipfontein was evident across South Africa, with a number of transportation planners in Johannesburg, eThekwini and Rustenburg in addition to Cape Town mentioning the experience as instrumental in ongoing BRT planning. Several suggested that the Klipfontein Corridor did not so much fail as morph into contemporary plans for BRT, both within Cape Town as well as across South Africa. The urgency to capitalize on this learning was further accelerated by South Africa's hosting of the 2010 Football World Cup, which resulted in various policy frameworks and legislation inviting cities to plan for integrated transportation networks. The funding scheme, the Public Transport and Infrastructure Systems Grant, was launched in March of 2005 to support investment in public transportation infrastructure, including planning, construction and improvements of new and existing systems. However, the decision to proceed with BRT rather than an alternative scheme, remained an independent decision of municipal policymakers. Those with prior experience with BRT "kept it alive" and "a lot of that did carry through to some of the projects that came to fruition in 2006", explained Lynne Pretorius [68]. Whereas local decisions regarding mobile ideas can be delayed by lack of funding, postponed because of politicking or stuck in bureaucracy, the learning is always ongoing.

These narratives illustrate the multiple temporalities of BRT learning, framing contemporary appreciation within the protracted nature of policy mobilities. Learning is a diffuse process and multiple examples are necessary to form policy models. Cynics dismissed Curitiba because it is a wealthy city with an exotic mayor and "people didn't think it could be replicated"; Quito was ignored because it is a small, historic city and "people didn't know about it" [27]; and Pereira did not have a charismatic mayor to travel around the world touting its BRT system. Bogotá is a major metropolis in a middle-income country and Transmilenio was a "game changer" [41]. BRT did not arrive in South Africa in 2006 as an unknown entity but rather as part of variations of good transportation planning and transit-oriented development, that were already present and been experimented with on several occasions. These seeds of BRT in South Africa substantiate the theoretical claims in this chapter that policy mobilities is a gradual, repetitive and delayed process.

Transportation Innovations Not Adopted

The multiple temporalities of BRT circulation emerged in a number of conversations with South African policy actors, illustrating their experiences with other transportation trends. Peck (2011a) raises alarm for the potentially faddish nature of emulation, claiming that entrepreneurial policy actors seem to be just as susceptible to fads as their corporate colleagues, reflecting growing anxieties for managing cities with limited financial and institutional capacity in an increasingly uncertain world. Bill Cameron, Director of Strategy and Implementation Monitoring for the National Department of Transport, compared BRT to other transportation fads, recalling, "Rail technology has been with us since the steam engine. It may have gotten faster but it is still the same technology. We had electric trams and trolleybuses. It was a fad to remove the trams all over the world but now we see that was a mistake" [81]. Several actors repeated his concerns that BRT may be just another policy fad, traveling through South Africa promising to be the urban panacea. Subethri Naidoo, Urban Governance Specialist at the World Bank, frames these arguments, "We jump from one policy fad to another without trying it and seeing how it works and that is for a variety of reasons" [25]. Her concern for the selective nature of urban transportation solutions relates to this discussion on the multiple temporalities of mobility, by suggesting that best practice innovations are often presented and then discarded for being erroneous and antiquated.

The arguments that follow consider these comments in reference to other similarly sensational interventions – several forms of personal rapid transit (PRT), which appeared in South Africa in the 1970s as the solution to the city's rapid suburbanization and road congestion; the monorail, which passed through South African cities alongside BRT and was also briefly hailed as the transportation innovation of the future contributing towards more livable cities; and the introduction of Gautrain, a high-speed rail service connecting Johannesburg and Tshwane, which opened in 2010. These demonstrate not only the faddish character of mobilities, but also the way in which processes of failed or stalled progressions contribute towards overall learning and exchange.

The first case reflects on the mobility of several forms of PRT, introduced in the 1970s across South African cities, and welcomed as a technological panacea. This technology, designed to carry between one and 100 people, promised to reduce the inefficiencies of public transportation. Because they move along a fixed monorail, are automatic, and only operate when there are passengers present, they were also sold as environmentally friendly. Variations of the technology arrived from Germany, Zimbabwe, Japan and the UK, but the idea never really caught on. The innovation was heavily promoted by Germany's Siemens who circulated the merits of H-Bahn overhead cabin transportation system, which carried 26 passengers in each cabin collecting and depositing them according to their desired destination. Japan exported another version of PRT, based on its achievements successfully moving over 50,000 passengers daily the 13 km between Tokyo's airport and the city center. Those cabins hang from a monorail to minimize the system's impact on the city. The cost for installing this system was R4 billion per kilometer. A third version also came from Germany, the Demag cabin taxi, which runs suspended from a guiderail carrying between four and 50 passengers. The UK designed yet another system called the Minitram, which could carry 14 people in each car with the ability to attach additional cars for a capacity of 900 passengers per hour. PRT technology was likewise promoted by the Zimbabwean Pneuways Company, who in 1977 introduced a design for a train that ran on rubber wheels along an overhead concrete rail, capable of carrying 30,000 seated passengers an hour that was expected to cost R1 million per kilometer. The system was proposed between Hillbrow, Braamfontein and the city center, but was deemed largely inappropriate for the suburbs. PRT was ultimately rejected by local transportation planners, who determined that the monolithic system was inappropriate for the vast distances across low densities in South African cities (*The Star* 1977).

Bruno Latour provides a colorful account of a long-drawn attempt to launch a PRT system in Paris, known as Aramis (Agencement en Rames Automatisées de Modules Indépendants dans les Stations, or in English, Automated Trains of Independent Modules in Stations). He recalls how PRT was fashionable in the 1970s when it seemed like "private cars were on their way out" and "we couldn't keep moving in the direction of mass transportation" (1996: 15). The innovation was popular because it served as a wistful reminder of the heyday of trams from the 1920s, which in spite of much chagrin, had been unceremoniously eradicated decades earlier (Anderson 1996). The realization of Aramis focused on the development of custom-designed system motors, sensors and other digital electronics, and most importantly, non-material couplings, which allowed the vehicles to move in unison without being physically attached. Latour, however, maintains that the challenges with Aramis surfaced from the political difficulties of convincing leaders of its merits, and had little to do with the technical feasibility of the project. Aramis never materialized for a number of reasons – concern for the safety of passengers traveling with strangers in private cars; lack of communication between developers and politicians; and possible misappropriation of funds which made the project prohibitively expensive. Despite the enthusiasm for PRT systems, to date, only three have been developed: a 13.2 km system built in 1975 in Morgantown, West Virginia; a 1.5 km track for intra-city transportation built in Masdar City, Abu Dhabi in 2010; and a 3.8 km path in London's Heathrow Airport, which in 2011 started carrying passengers from the car park to the airport. This example of another transportation technology illustrates the lengthy nature of policy conversations, in which innovation is sometimes repeatedly ignored or considered inappropriate, and at other times merely takes longer to catch on. Unlike BRT, however, which was eventually embraced, this case of PRT provides insight into the way in which certain innovations are never approved regardless of the number of attempts.

Another example of failed transfer of transportation innovation is the monorail, which moved through South African cities at the same time as BRT. Like PRT, the monorail was briefly hailed as the transportation innovation of the future contributing to more livable cities. Cape Town, eThekwini and Johannesburg toyed with the idea of building a monorail between the airport and town, along the beachfront and between Soweto and town, respectively, with systems staunchly promoted by industry and for the most part, opposed by transportation experts [8; 17; 50]. A number of South African planners and politicians recalled the attempted construction of a monorail as "one of our biggest nightmares" because "we would still be building it" and "we would be in debt forever" [8]. In Johannesburg, the 2007 "monorail disaster" [31] was driven by a Malaysian consortium who agreed to build, finance and operate a R12 billion monorail through Soweto from Protea to the Bree Street taxi rank in town. In exchange, the consortium would receive a 30-year concession for the land around the monorail (Moodley 2007). This model is essentially a development scheme, which raises concerns for the quality of the service and the type and location of development along the route [39]. Gauteng Province initially approved the project without sufficient stakeholder engagement or adequate due diligence, but because new legislation prevents provinces from making transportation decisions for roads within municipal boundaries, the agreement was nullified [38]. Similar models were considered in Cape Town and eThekwini but without any agreements or commitments.

South African policy actors are aware of global transportation achievements and are sensitive to the evolution of chic solutions, especially in reference to their own transportation trends. Madeleine Costanza, who worked for IIEC from 1998 to 2002, compares the approval of new transportation approaches to uptake of "a new cellphone or new fashion". Costanza states, "Generally, people make incremental changes. We think we know what we like in fashion, we think we know what to do with our phones, until we meet a friend or colleague who does something dramatically different – and we like that difference – we probably have never thought about that difference before" [23; 30]. Transportation innovations are by nature faddish suggested Gershwin Fortune, Manager of System, Planning and Modeling for BRT in Cape Town from 2007 to 2016, and he therefore contends that transportation planners build from their experiences with previous transportation fashions to evaluate new trends. He argues that this history gives South Africa a technical and political edge when localizing the BRT model [56]. The monorail was such an obvious instance of "policy colonization" [8], in which the concept was promoted vociferously by international property developers, who promise to build the monorail at a low cost in exchange for a large percentage of the profits, that municipal planners immediately dismissed it as part of a political agenda rather than a technological or social marvel.

Another transportation project is the Gautrain, a rapid railway connecting Johannesburg with Tshwane, as well as within the intermediate suburban nodes and O. R. Tambo International Airport. The 80 km route and 10 stations cost roughly R30 billion, with a projected ridership of 120,000 passengers per day. The project was intended to reduce traffic congestion on the N1/Ben Schoeman Freeway between Johannesburg and Tshwane, to stimulate economic growth by directly creating jobs related to the construction and operation of the train, and more generally, to promote tourism and to encourage environmental-friendly methods of transportation. Construction began in February 2006, although planning began many years earlier. As early as 1997, then-Premier of Gauteng Province, Tokyo Sexwale publicized an exploration into the possibility of introducing a rapid railway in the region; in 2000, then-Premier Mbhazima Shilowa broadcast that the Province would build the first high-speed interurban transportation system on the African continent; and it took another two years before commencing an environment impact assessment for the project (Thomas 2013). It is certainly worth noting that South Africa was awarded the opportunity to host the 2010 World Cup in March of 2005 and by December, the South African government had authorized the project to proceed. The initial track was operational in June 2010, although the remaining sections opened after the World Cup in August 2011, and the final section between Rosebank and Johannesburg Park Station, which was delayed due to water seepage in the tunnel, opened in June 2012.

The story of the Gautrain provides evidence that BRT was not the only influential transportation idea moving through South African cities. Here was a robust model for a new rail project realized after a long planning period that also changed the perception of public transportation in South Africa. Gautrain was part of the same idea of a comprehensive intervention that took people out of their cars but, whereas BRT provides improved services for the poorest residents, Gautrain is often criticized for servicing the wealthier, northern suburbs of Johannesburg and pricing tickets too high for the average South African commuter [88]. Jeremy Cronin, Deputy Minister of the National Department of Transport from 2009 to 2012, describes the Gautrain as a "great public transport project but the wrong priorities". Because of those experiences, Cronin reckons, cities learned how to plan, design and construct major transportation interventions [62]. Cities also learned that sometimes building a single transit corridor is not a far-reaching transportation solution leading Ibrahim Seedat, Director of Public Transport Strategy at the National Department of Transport, to push Johannesburg to continue building BRT in spite of low ridership, which threatens the viability of the system [38]. These experiences are indicative of the many different experiences that facilitate policy flows, and the way in which learning experiences are cumulative, shaping future decisions.

With so many transportation solutions moving through South African cities, the issue of policy fads emerged, with a number of South African respondents concerned that BRT may be just another policy fad flowing through South Africa promising to be the urban panacea. "Africa was the flavor-of-the-year but now Southeast Asia is the flavor-of-the-year so the money has gone there", explains Pauline Froschauer, "It is impossible to get money for projects in Africa now when it is not trendy, but everyone will throw money in Southeast Asia" [75]. Several policy actors argued against the postulation that BRT is just the latest craze, instead maintaining that BRT is the best solution for the South African city. Andre Frieslaar, Director of BRT Planning, Traffic Engineering and Transport Planning at HHO Africa in Cape Town understands BRT to be a long-term solution. "I think BRT will be around for a while. It is a very attractive technology for Africa and Asia, and I think it has huge benefits, transformational opportunities. At this stage, it is clearly not a fad but has its own life, but what will make it survive will be coming up with more affordable ways of running the system" [86]. Concerns for financial viability were often made in reference to the Gautrain, a high-speed rail service connecting Johannesburg and Tshwane. Sam Zimmerman, Senior Urban Transport Specialist at the World Bank argues that BRT will continue to be the best solution because it is more affordable than other previously initiated solutions. "I think BRT will continue. The trend will grow if only because people want transportation systems" [33].

Conclusion

This chapter attends to the varying temporalities of policy mobility, by demonstrating that BRT circulation is a more convoluted and long-lasting process than ordinarily considered. My arguments move beyond the literature on fast policy, which excuse policy transfer as a "clumsy form of crisis displacement" (Peck 2002: 350), instead reflecting on the multiple temporalities of mobility and the need for serial introductions before the innovations take root locally. In the modern era, knowledge of international policies may be exchanged more frequently through additional fact-finding trips, interpersonal contacts and a greater role for international advocacy but acts of sharing innovation do not automatically lead to uptake and shortened policy cycles. Constant reminders normalize the innovation within the policy environment thereby making it appear as the only available solution. While it may seem as if pre-approved policies shorten the gestation time from policy introduction to policy adoption, gradual and repetitive attempts to apply diffusive innovations ensure that when the context is ready, the turnover time appears accelerated. This more exhaustive exploration of the circulation process also offers an opportunity to depart from the theoretical discussions of policy mobility as a contemporary phenomenon of fast policymaking, instead exposing a history of protracted and idiomatic learning.

The history of policy innovations and its influence on BRT endorsement, adds a critical layer to the scholarly literature on policy movement. In the first place, just as in the past, practices from elsewhere are embraced because of the particularities of the local policy actors and sociopolitical circumstances that shape each individual instance of learning. Slower processes of adoption may result from careful considerations of alternatives, financial limitations and institutional politicking, all of which are firmly grounded in local circumstances. Greater understanding of the context of mobility and the role of local conditions in the localization of international best practice, adds a counterargument to the suggestion that policy mobilities is characterized by the pervasive influx of supply-side solutions, usually introduced by policy experts from outside the locality, who create, package and sell prototypes as best practice. Rather, a historical approach diminishes the power of these international experts and turns a spotlight on the local policymakers who approve (and modify) global best practice.

The final theme emerging from a review of public transportation experimentation reflects on the role of the specific innovation, and the manner in which policy ideas mutate and transform through transfer. In this instance, variations of BRT appeared several times in South Africa before being introduced in the urban terrain. The Bogotá model was finally mimicked because in addition to its technical accomplishments, it also includes a mechanism for incorporating existing transit operators, an improvement over the Curitiba model. This sociopolitical feature made it attractive to cities around the world wrestling with informal transport systems and was especially necessary to attract support in South Africa (see the "BRT and Taxi Transformation" section). It seems that policy ideas change as they travel, but this mutation does not only happen across geographical space but also over time. Greater understanding of the multiple temporalities of learning explains why BRT was adopted rather than alternative forms of urban transportation.

The patterns of innovation and experimentation outlined in the story of BRT circulation illustrate the need for persistent introduction and alteration before transfer can occur. Certainly, the mobility of global ideas can be a difficult process, and thus time is a critical factor in policy adoption. Innovations like BRT have spread across the globe, not via a process of fast policy, but through a form of hegemonic power that relies on repeated suggestions that ultimately designate something as a best practice to local policy implementers. These arguments make an important contribution to the existing literature on policy mobilities, by considering the way in which best practice flows through a more subtle and consistent method of persuasion.

Chapter Seven
Conclusion

Introduction

The adoption of BRT in South Africa is the product of extensive dialogue and exchange between South Africans and South Americans, and among South African architects, urban designers, financial managers, transportation engineers, politicians and other officials. Learning took place through conferences, study visits and a lengthy consultation process with existing transit operators. The genesis of this process goes back to July 2006 when international experts presented the attributes of BRT at a workshop at the Southern African Transport Conference, and their subsequent visits in August 2006 to Cape Town, eThekwini, Johannesburg and Tshwane; and in many ways, BRT adoption began when Johannesburg City Council became the first South African city to approve BRT in November 2006. What emerged was a formal transportation service managed by former taxi operators with the support of local government. The impact of BRT in addressing issues of accessibility, density and inequality in a unified South African city remains to be seen.

South Africans proudly boast of an association between their BRT systems and those in Bogotá and Curitiba, as well as Quito, Guayaquil, Pereira, Santiago, and Sao Paulo (e.g. City of Johannesburg 2011). They particularly call upon the experiences of policymakers in Bogotá to defend their determination to improve the existing transportation system, in spite of initial staunch opposition from national government and minibus taxi operators (see the "The International Context of BRT Circulation" section). While it may appear to have been a relatively

How Cities Learn: Tracing Bus Rapid Transit in South Africa, First Edition. Astrid Wood.
© 2022 Royal Geographical Society (with the Institute of British Geographers). Published 2022 by John Wiley & Sons Ltd.

straightforward process in which cities learned of and implemented BRT in a timely and efficient manner, this book demonstrates that policy mobilities and adoption is gradual, repetitive, delayed, and always subject to local political contestation. Whereas Johannesburg's Rea Vaya opened just three years after learning of the Bogotá model of BRT, Nelson Mandela Bay's Libhongolethu and Tshwane's A Re Yeng stalled. Moreover, while Rustenburg has made a considerable effort to understand the experiences of BRT implementation and operations in Cape Town and Johannesburg, eThekwini has made significantly less effort to learn from its compatriots. The chapters in this book reconsider the role of local policymakers and their experiences of adopting BRT across South African cities.

This concluding chapter reviews the book's most important arguments. In the next section, I summarize how cities learn. I outline my main contributions to the policy mobilities literature by interrogating the local interpretation of BRT, the various technical and political leaders shaping it, the inter-urban and intra-urban politics among South African cities, and the particular history of experimentation that has taken place. In the third section of this chapter, I conclude by reflecting on the impact of BRT on the political and spatial landscape of the South African city.

Reflecting on How Cities Learn

The aim of *How Cities Learn* is to trace the mobility of BRT from Bogotá to South Africa in order to understand why it was adopted by South African cities. The book unravels the story of BRT adoption by drawing on insights from the policy mobilities literature to understand how local factors shape a determination to accept or reject mobile knowledge. This approach brings adoption to the fore. I argue that in order to understand how and why policies are adopted from elsewhere, additional theorization is needed of the particularities of the policy, people and place, which make it timely and appropriate. The key question in *How Cities Learn* is not whether policies transfer adequately or to assess the achievements of a mobile policy, but rather how traveling policies are interpreted and applied within the adopting locality.

Chapter 3 is the first chapter to address this question. It considers the nuanced characteristics of BRT and Bogotá's Transmilenio, before honing in on the particular features of BRT in South African cities. Chapter 3 discloses that as BRT travelled around the world, it was not simply mimicked but rather deployed elsewhere anew, reflecting the conditions in each new city. This argument reveals the way in which policy adoption is not a process of cloning but rather involves practices of fragmentary borrowing and of emulation, imitation and repetition of lessons and best practices. BRT does not circulate as an established set of blueprints or plans guiding local policymaking, but as a practice that captures some essence of the achievements of the original version. In so doing, Chapter 3 explains how South African policymakers interpreted and applied the BRT concept, looking specifically at the station platform, bus, bus lane, and routing. By unraveling the application of BRT in South Africa, I illustrate that BRT circulated because of its flexibility, which allowed it to insert itself into complex political conditions that differed considerably from the original context.

This understanding of BRT as a set of guidelines and recommendations rather than prescriptive model ready for duplication, is illustrated by the challenges South African cities faced with existing taxi operators when introducing BRT. This became the most prominent challenge for BRT because it was highly politicized. During the implementation of Rea Vaya in 2009, President Jacob Zuma ordered the city to halt construction until a resolution could be reached with the taxi operators, and in October 2009, four armed assailants attacked Johannesburg city councilor and BRT proponent Rehana Moosajee. The assailants were never identified and it is widely believed that the shooting was a warning from the taxi industry. Challenges such as these are the basis for my argument that it is sometimes difficult to introduce a policy into another context. I examine the institutional opposition from the existing operators and the ways in which South African policy actors molded BRT to fit within a number of previous attempts at reforming the minibus taxi industry. This examination exposes a more complicated and nuanced view of the story of BRT in South Africa, with objections now long forgotten because of the realization of the systems in Cape Town and Johannesburg.

Chapter 4 reinterprets the process of assembling, mobilizing and adopting BRT through the role of global and local policy actors and their associations. This chapter analyzes the variety of state and non-state, local and global, professional and nonprofessional actors sharing, translating, mobilizing, adopting, and implementing BRT across South African cities. Each policy actor has their own shifting and overlapping role: policy mobilizers introduce best practice policy models, intermediaries connect these actors and their knowledge with local policymakers, and local policymakers adopt (or reject) the innovation. The demand for readily consumable models reveals more about the needs of the consumers than the efficacy of the model itself. In many instances, after completing their term in office, local policymakers travel great distances to disseminate their achievements to eager ears at paid speaking engagements where their investigations are met by extrospective actors who actively monitor narratives of success. I reason that this process is also the outcome of contracting city budgets or staff reductions, which put pressure on remaining personnel to promptly produce proven successes that bring employment and investment opportunities back into the city. BRT adoption, which involves personalities, politics and power, is thus far more complex than a simple supply-and-demand process.

The second part of Chapter 4 focuses on the associations, collaborations and networks of international and South African policy actors driving the circulation and adoption of BRT. I reason that cities learn of circulated forms of knowledge through innovative and dynamic individuals, moving best practice across their formal and informal networks, yet policy is only adopted through local actors with intergovernmental support. Contrary to the usual understanding of BRT as an externally driven process, in this instance South African policy actors invited BRT. In many cases, however, the international teams were essential in moving the process along by providing technical advice as to how to construct and operate a BRT. These arguments are critical within discussions regarding the directionality of policy flows. Disentangling the diversity of South African engagements with policymakers around the world, as well those within South Africa, adds a critical dimension to understandings of policy mobilities by considering the way in which best practice flows through a more subtle and consistent method of persuasion. It also furthers my understanding of the translocal assemblages, composed of local and international, professional and voluntary, government and nongovernment actors and associations that further notions like BRT.

Chapter 5 relocates the story of BRT adoption within the local politics guiding BRT uptake, by focusing on the exchanges between exporting and importing sites as well as across South African localities and within each city. The first argument in this chapter utilizes the story of BRT adoption in South Africa to unravel the role of south–south connections in local policymaking. I demonstrate that circulated policies are adopted because of their connections with elsewhere, embodied through political associations and representational power, which make them mobile and applicable. The reproduction of best practice then is never a straightforward technical process or a rational survey of best practices, but rather a political process in which local policymakers deploy different meanings to justify their policy decisions. In this case, South African policymakers first utilize the association with Bogotá to rationalize the decision to implement BRT and then employ hierarchical notions to ignore best practices taking place elsewhere. Illustrating the way in which policymakers deploy different meanings of the global south to justify local policy decisions adds a critical dimension to understandings of comparative urbanism. This chapter contributed to the policy mobilities field by demonstrating that adopting localities are instrumental in cultivating a receptive ground for the application of mobile policy. It also provides an example of how to research transnational policy flows through the adopting locality.

The second part of Chapter 5, repositions the discussion within South African cities and their relationships. I interpret the different reactions to BRT as a feature of the sociopolitical variations across the six South African cites, arguing that these relational and territorial distinctions lead to discrepancies in uptake. These variations include a range of structural and non-structural issues including economic competitiveness, geographical size and financial capacity, as well as the political perspicacity of particular policy actors and associations in promoting circulated policies. The South African example reveals that competitive tendencies shape both political and technical interactions. It is important to consider whether a city adopted BRT to illustrate the superiority of a competing political party, or because it aligned with its global aspirations, or merely because it was easy to fund a major infrastructure investment. This competitiveness shaped the kind of system each city implemented. It need not be a hindrance to BRT adoption. In fact, competitive tendencies accelerated the implementation of BRT by motivating South African policy actors to introduce BRT in spite of obstacles, such as hostility from existing taxi operators. In South Africa, having an operational BRT system became an indicator of progress and innovation. No wonder so many South African cities were drawn towards BRT.

While it may appear as if South African cities promptly applied BRT once they learned of it, Chapter 6 reveals that these processes of learning took place over multiple periods during which policy actors ruminate and consider best practice, perhaps shifting the model to suit local conditions. Comparable iterations of BRT were proposed several times across South African cities before 2006 without taking root. These earlier encounters, which include the first published discussion of BRT in South Africa in a 1973 conference report, study visits to Curitiba in the 1990s and a failed attempt to implement a Bogotá-style BRT system in Cape Town in 2003, were instrumental in creating a fertile ground for later adoption practices. It is also important to recognize that there were a number of other solutions also presented as optimal that sometimes spread, sometimes were ignored and sometimes rejected in accordance with local conditions. Many ideas move through the urban terrain, any number of which are not introduced and many others are swiftly forgotten. Chapter 6 considers the multiple temporalities through which policies flow and the outcome of policy failure in the process of policy mobilities. This chapter illustrates the progressive nature of learning, through which previous experiences was instrumental in ongoing decisions to adopt and implement BRT. In unraveling the story of BRT adoption, Chapter 6 departs from the theoretical discussions of the policy mobilities process as a rapid phenomenon, instead demonstrating it as gradual, repetitive and delayed. This chapter is not merely a historical overview, but rather an attempt to understand the various people and processes assembled and mobilized from the perspective of the adopted model. Such an intellectual endeavor expands our understanding of how cities innovate, and contributes to the literature by revealing the constant and persistent nature of policy mobilities.

Each of these arguments accumulates support for the critical role of localities in influencing policy adoption. These findings are not only useful within South Africa but also provide guidance to cities around the world. Whereas previous accounts focus on the attributes of BRT, comparing its speed and efficiency to a rail service, and its cheap and easy implementation to a bus service, *How Cities Learn* reveals how localities interpret, frame and structure circulated forms of knowledge, forming and fashioning it to suit prevailing geographical and sociopolitical conditions and connections.

Reflecting on BRT in South Africa

Ideally, the Rea Vaya and MyCiTi bus follows a schedule: passengers prepay using cashless cards, travel information is clearly posted at purpose-built glass and concrete stations, and journey times are reduced by moving through demarcated busways that ensure predictable travel speeds. Implementing each of these services – developing a schedule to adequately service peak and off-peak demand, building partnerships with South African banks to support the travel card, designing the stations in partnership with community input, consulting with minibus taxi operators, and enticing the public onto the bus – required extensive efforts by local and international policy actors. In South Africa, however, BRT did not materialize as imagined.

Fifteen years since BRT first arrived at that 2006 workshop, construction is still ongoing. A second phase of BRT is now operational in Johannesburg, Cape Town and Tshwane; and other South African cities such as Ekurhuleni continue to plan and develop systems. Rea Vaya Phase 1C and MyCiTi Phase 2 are years behind schedule.

Ridership on the operational BRT systems has been far lower than anticipated. Each city's BRT is expected to carry at least 1 percent of the city's population. In 2019, however, on average BRT carried just 0.7 percent of each city's population (0.95 percent in Johannesburg, 0.24 percent in Tshwane and 0.13 percent in Ekurhuleni). Only Cape Town (1.48 percent) exceeds this threshold (National Treasury 2019). South Africans rationalize that buses are not used because they are not available, infrequent and generally inaccessible. In Gauteng, bus travel has declined from 267,066 trips per day in 2014 to just 59,408 in 2019 or just 2 percent of daily trips. By contrast, driving as a main mode of transportation has increased from 22 percent in 2014 to 27 percent in 2019. The main reasons for not using buses according to the 2019/20 Gauteng Household Travel Survey were that "no bus available" (45 percent), "not available often enough" (11 percent), "buses don't go where needed" (9 percent) or "bus stop too far from destination" (5 percent). Figures that might have featured more prominently are that the "bus too expensive" (3 percent), "travel time too long or too slow" (0.1 percent), "have to change transport" (0.1 percent) or "prefer taxi" (0.2 percent) (Gauteng Department of Roads and Transport 2020).

In addition, the financial viability of BRT in South Africa remains a challenge. BRT fares were expected to cover at least 35 percent of operational costs. Of those cities with operational BRT systems, only about 30 percent of costs (41.1 percent in Cape Town, 16.8 percent in Ekurhuleni, 38.5 percent in Johannesburg and 21.5 percent in Tshwane) are covered by the ridership fares (National Treasury 2019). Over the long-term BRT operations will need to be subsidized, an outcome BRT promoters promised would never happen.

Managing BRT operations has been an issue for South African cities. Rea Vaya and MyCiTi have been riddled with bus strikes and vandalism to stations. Both Cape Town (in 2012) and Gauteng (in 2020) experimented (to varying degrees of success) with the establishment of a transportation authority. The Cape Town Transport Authority (TCT) was tasked with integrating the city's transportation system by amalgamating competing public transportation modes and coordinating planning between national, provincial, and local spheres of government. In 2016, Cape Town extended the functions of TCT to include urban development (urban planning, human settlements and sustainability) alongside integrated transport and, in 2017, the Transport and Urban Development Authority (TDA) became the city's transportation authority (Dentlinger 2016). The TDA's extended responsibilities were expected to help the city achieve the Transit Oriented Development Strategic Framework (City of Cape Town 2016). These additional urban planning and human settlement functions proved too challenging for TDA, and in May 2019, TDA was disbanded.

In Johannesburg, the City attempted (somewhat unsuccessfully) to use BRT to spark development along the bus corridors, improve transportation connections to previously marginalized areas of the city, and densify the area around the stations. In May 2013, shortly after the launch of Rea Vaya Phase 1B, Johannesburg's then-Mayor Parks Tau launched an ambitious program of transit-oriented development, dubbed the "Corridors of Freedom" (Tau 2013). The plan promised to reshape the city by linking transportation interchanges with high-density, mixed-use residential, retail and business developments as well as educational and leisure facilities. Interventions included clinics, libraries, parks, sports facilities, cycleways and pedestrian crossings. The Corridors of Freedom, which had been under-resourced throughout, by 2018, was defunded. Similar policies of transit-oriented development have also been developed in Cape Town (City of Cape Town 2016) as well as elsewhere in South Africa (Cooke et al. 2018), with similarly scanty results.

The long-term outcome of BRT in South Africa remains to be fully understood. These preliminary reflections suggest a range of challenges associated with policy change. Indeed, policymaking like policy mobilities is never complete but always in motion, constantly moving and mustering, touching-down and taking-off, entangling with politics and disseminating the involvement of elsewhere around the globe. Greater understanding of how cities learn, I hope, will inspire policymakers to carefully weigh the opportunities presented through their engagement with best practices.

Appendix A: Interview Schedule

Ref. number	Title	Organization	Place of interview	Date of interview
1	Executive Manager: Programs	South African Cities Network	Johannesburg	23-Jan-12
2	Executive Manager: Planning and Strategy	Johannesburg Development Agency	Johannesburg	24-Jan-12
3	Program Manager	South African Cities Network	Johannesburg	26-Jan-12
4	Professor: Planning	University of Witwatersrand	Johannesburg	30-Jan-12
5	Transportation Planner	Johannesburg Road Agency	Johannesburg	31-Jan-12
6	Acting Executive Director: Strategy, Policy and Research; Chairperson of PlanAct Board	South African Local Government Association	Johannesburg	1-Feb-12
7	Political Liaison	South African Cities Network	Johannesburg	1-Feb-12
8	Director: BRT Implementation	City of Johannesburg	Johannesburg	3-Feb-12
9	Deputy Executive Director	Human Sciences Research Council	Cape Town	6-Feb-12
10	Professor	African Centre for Cities	Cape Town	7-Feb-12

How Cities Learn: Tracing Bus Rapid Transit in South Africa, First Edition. Astrid Wood.
© 2022 Royal Geographical Society (with the Institute of British Geographers). Published 2022 by John Wiley & Sons Ltd.

APPENDIX A: INTERVIEW SCHEDULE

Ref. number	Title	Organization	Place of interview	Date of interview
11	Director: Centre for Transport Studies	University of Cape Town	Cape Town	7-Feb-12
12	Researcher	University of Cape Town	Cape Town	7-Feb-12
13	Researcher	African Centre for Cities	Cape Town	7-Feb-12
14	Cabinet Member	Parliament	Cape Town	7-Feb-12
15	Chairperson	National Executive Committee of the ANC	Cape Town	7-Feb-12
16	Managing Director	Organization Development Africa (ODA) South Africa	Cape Town	8-Feb-12
17	Chief Executive Officer	Cape Town Partnerships	Cape Town	13-Feb-12
18	Program Manager	Organization Development Africa (ODA) South Africa	Cape Town	13-Feb-12
19	Policy Analyst	Development Bank of Southern Africa	Johannesburg	16-Feb-12
20	Transport Specialist	Development Bank of Southern Africa	Johannesburg	16-Feb-12
21	Transport Specialist Advisory Unit	Development Bank of Southern Africa	Johannesburg	16-Feb-12
22	Member of the Mayoral Committee for Transport	City of Johannesburg	Johannesburg	17-Feb-12
23	Director	International Institute for Energy Conservation	Vancouver, Canada	19-Feb-12
24	Knowledge Management Analyst	Cities Alliance, World Bank	Tshwane	20-Feb-12
25	Director: City Support Group	World Bank Institute	Tshwane	20-Feb-12
26	Executive Director	Victoria Transport Policy Institute	Vancouver, Canada	20-Feb-12
27	Executive Director	VivaCities	Manila, Philippines	23-Feb-12

Ref. number	Title	Organization	Place of interview	Date of interview
28	Coordinator: Participatory Governance Program	PlanAct	Johannesburg	24-Feb-12
29	Chief Operations Officer	Johannesburg Development Agency	Johannesburg	28-Feb-12
30	Director	International Institute for Energy Conservation	Vancouver, Canada	28-Feb-12
31	Director of Research	Gauteng City Region Observatory	Johannesburg	29-Feb-12
32	Senior Transport Economist	World Bank	Washington DC, USA	1-Mar-12
33	Senior Urban Transport Specialist	World Bank	Washington DC, USA	1-Mar-12
34	Architect	Ikemeleng/ Osmond Lange Architects	Johannesburg	2-Mar-12
35	Secretariat	National Planning Commission	Tshwane	5-Mar-12
36	Chief Executive Officer	Johannesburg Development Agency	Johannesburg	6-Mar-12
37	Deputy Director: Transport	City of Tshwane	Tshwane	6-Mar-12
38	Director: Public Transport Strategy	National Department of Transport	Tshwane	9-Mar-12
39	Chief Director, Public Finance: Urban Development and Infrastructure	Department of Treasury: Transport	Tshwane	12-Mar-12
40	Head	Gauteng Planning Commission	Johannesburg	13-Mar-12
41	Executive Director	Transportation Alternatives	New York, USA	14-Mar-12
42	Director: Traffic Engineering and Operations Transport & Roads Department, Roads & Stormwater Division	City of Tshwane	Tshwane	14-Mar-12

Ref. number	Title	Organization	Place of interview	Date of interview
43	Head: Corporate Strategy and Policy	City of Johannesburg	Johannesburg	14-Mar-12
44	Professor	University of Witwatersrand	Johannesburg	15-Mar-12
45	Senior Manager: Municipal Institute of Learning	eThekwini Municipality	eThekwini	19-Mar-12
46	Head: Development Planning, Environment & Management	eThekwini Municipality	eThekwini	19-Mar-12
47	Deputy Head: Strategic Transport Planning	eThekwini Transport Authority	eThekwini	20-Mar-12
48	Transport Historian	eThekwini Transport Authority	eThekwini	20-Mar-12
49	Head: eThekwini Transport Authority	eThekwini Transport Authority	eThekwini	22-Mar-12
50	Head: International and Governance Relations	eThekwini Municipality	eThekwini	22-Mar-12
51	Director	Luis Willumsen Consultancy	London, UK	23-Mar-12
52	Lecturer	Universität Erlangen-Nürnberg	Johannesburg	28-Mar-12
53	Principal Urban Designer	Albonico Sack Mzumara Architects	Johannesburg	30-Mar-12
54	Director	Hunter van Ryneveld	Cape Town	3-Apr-12
55	IRT Manager Infrastructure and Development	City of Cape Town	Cape Town	3-Apr-12
56	Manager: System, Planning and Modeling	City of Cape Town	Cape Town	3-Apr-12
57	Manager: IRT Integration & Project Management	City of Cape Town	Cape Town	3-Apr-12
58	Chief Executive Officer	Garden Cities NPC	Cape Town	4-Apr-12
59	Researcher	University of Cape Town	Cape Town	4-Apr-12
60	Director	Strategies for Change	Cape Town	5-Apr-12
61	Director	ITS Ltd	Cape Town	5-Apr-12

APPENDIX A: INTERVIEW SCHEDULE

Ref. number	Title	Organization	Place of interview	Date of interview
62	Deputy Minister	National Department of Transport	Cape Town	5-Apr-12
63	IRT Manager Infrastructure and Development	City of Cape Town	Cape Town	6-Apr-12
64	Director	Transport Futures	Cape Town	9-Apr-12
65	Member of the Mayoral Committee for Transport, Roads & Stormwater	City of Cape Town	Cape Town	10-Apr-12
66	Director	African Centre for Cities	Cape Town	11-Apr-12
67	Professor	University of Cape Town	Cape Town	11-Apr-12
68	Director	Pendulum Consultants	Cape Town	11-Apr-12
69	Director	Bicycle Empowerment Network	Cape Town	12-Apr-12
70	Director of Transport	City of Cape Town	Cape Town	13-Apr-12
71	Program Coordinator: Africa Program	Sustainable Cities International Network	Dar es Salaam, Tanzania	22-Apr-12
72	Consultant	Institute for Transportation and Development Policy (ITDP)	Bogotá, Colombia	27-Apr-12
73	Executive Director	Institute for Transportation and Development Policy (ITDP)	New York, USA	28-Apr-12
74	Chairperson: Dept. of Transport & Supply Chain Management	University of Johannesburg	Johannesburg	2-May-12
75	Project Manager: Rustenburg Integrated Network Joint Venture	Namela Ltd	Tshwane	3-May-12
76	Project Manager: Urban Strategic Planning Committee	United Cities and Local Governments	Barcelona, Spain	7-May-12

Ref. number	Title	Organization	Place of interview	Date of interview
77	Mayor	City of Cape Town	Cape Town	7-May-12
78	City Manager	eThekwini Municipality	eThekwini	11-May-12
79	Executive Director	8-80 Cities	Johannesburg	16-May-12
80	Member of the Mayoral Committee for Transport	City of Johannesburg	Johannesburg	16-May-12
81	Director: Strategy and Implementation Monitoring	National Department of Transport	Tshwane	17-May-12
82	Director: City Support Program	National Treasury	Tshwane	17-May-12
83	Program Manager: City Support Program	National Treasury	Tshwane	17-May-12
84	Director	Black Earth Consulting	Johannesburg	29-May-12
85	Director	Designing South Africa	Cape Town	5-Jun-12
86	Director: Bus Rapid Transit Planning, Traffic Engineering & Transport Planning	HHO Africa	Cape Town	5-Jun-12
87	Director: Engineering	HHO Africa	Cape Town	5-Jun-12
88	Editor	Mobility Magazine	Cape Town	25-Jun-12
89	Executive Director: Transport	City of Cape Town	Cape Town	3-Jul-12
90	Consultant	Disability Solutions	Johannesburg	12-Jul-12
91	Chief Executive Officer	South African Cities Network	Johannesburg	13-Jul-12
92	Corporate Director/ Vice President	Putco/Southern African Bus Operators Association	Johannesburg	13-Jul-12
93	Senior Associate	Arcus-Gibb	Johannesburg	17-Jul-12

Ref. number	Title	Organization	Place of interview	Date of interview
94	Chairperson/Director: Corporate Affairs and Communications	Greater Johannesburg Regional Taxi Association/PioTrans	Johannesburg	17-Jul-12
95	Specialist: Strategy and Action Plan Development	TransForum Business Development	Tshwane	18-Jul-12

Appendix B: Features of BRT systems in South Africa

Name	*Rea Vaya* Phase 1A	*Rea Vaya* Phase 1B	*Rea Vaya* Phase 1C	*MyCiTi* Phase 1A	*MyCiTi* Phase 1B	*MyCiTi* Phase 2
Location	City of Johannesburg (JNB)			City of Cape Town (CPT)		
Population (ranking within South Africa)	4.5 million (1st)			3.8 million (2nd)		
Name	"We are moving" in Sotho			-		
Common name	Bus Rapid Transit			Integrated Rapid Transit		
Launch date	Aug-09	Oct-13	Sep-16	May-11	November 2013; April 2014	Jul-14
Building commences	Oct-07		Mar-14	Sep-08		Feb-14
Stage	operational	operational	construction of 1C and preliminary design for Phase 2	operational	operational	preliminary construction for Phase 2

How Cities Learn: Tracing Bus Rapid Transit in South Africa, First Edition. Astrid Wood.
© 2022 Royal Geographical Society (with the Institute of British Geographers). Published 2022 by John Wiley & Sons Ltd.

A Re Yeng Inception Phase	A Re Yeng (entire network)	Yarona Phase 1	Yarona (entire network)	Go Durban! Phase 1	Libhongolethu
Tshwane Metropolitan Municipality (PTA)		Rustenburg Local Municipality (FARG)		eThekwini Metropolitan Municipality (DBN)	Nelson Mandela Bay Metropolitan Municipality (NMBM)
3 million (5th)		550,000		3.44 million (3rd)	1.2 million (6th)
"Let's go" in Sotho		"It is ours" in Seswana		–	"Our pride" in Xhosa
Tshwane Rapid Transit		Rustenburg Rapid Transit		Integrated Rapid Public Transport Network	Integrated Public Transport System
Jul-14	Mar-16	Jan-16	2020	2018 (entire system in 2027)	January - November 2013
Jul-12	Jun-13	Jun-12	–	Mar-14	May-07
final stages of construction on inception phase		midway through construction of starter service		final stages of preliminary design; early stages of construction	pilot program completed in November 2013 and now in state of postponement

(Continued)

Name	Rea Vaya Phase 1A	Rea Vaya Phase 1B	Rea Vaya Phase 1C	MyCiTi Phase 1A	MyCiTi Phase 1B	MyCiTi Phase 2
Services operating	1 trunk route; 3 complementary routes; 5 feeder services	1 trunk route; 2 complementary routes; 5 feeder services	1 trunk route; 3 complementary routes; 2 feeder services	2 trunk services; 9 feeder services	Dunoon to Montague Gardens and Century City, with attached feeder routes	Metro South East, including Khayelitsha and Mitchells Plain
Modal split (private:public transport - bus, rail, taxi)	53:47 - 9% bus, 14% rail, 72% taxi; 5% other			50:50 - 10% bus, 60% rail, 30% taxi		
Management agency	City Transport Department (Contracted Services Unit)			City of Cape Town Department of Transport		
Number of staff in project office	80			100+		
Implementing agency	Johannesburg Development Agency			City of Cape Town Department of Transport		Transport for Cape Town (TCT) formed October 2013

A Re Yeng Inception Phase	A Re Yeng (entire network)	Yarona Phase 1	Yarona (entire network)	Go Durban! Phase 1	Libhongolethu
1 trunk and 4 feeder routes - CBD (Paul Kruger Street) to Hatfield	2 routes from Kopanong via CBD to Mamelodi	12 (2 trunk; 1 direct; 9 feeder); 240 bus stops	51 (6 trunk; 19 direct; 26 feeder)	3 BRT routes and 1 rail corridor (C1 Bridge City to Durban CBD, C3 Bridge City to Pinetown, C9 Bridge City to Umhlanga Corridor and the rail corridor: C2: Bridge City and KwaMashu via Berea Road to Umlazi and Isipingo)	five routes: the Khulani Corridor, Kwazakhele, Motherwell and New Brighton to the city centre via Korsten
	66:44 - 8% bus, 7% trains, 17% taxis	30:70 - 10% bus, 80% taxi, 10% other - 84% of Rustenburg residents do not own a car		52:48 - 18% bus, 30% taxi, 6% rail, 46% private car	49: private; 51% public
	City of Tshwane IRPTN/BRT Unit	Rustenburg Integrated Network joint venture		eThekwini Transport Authority	Royal HaskoningDHV
	25	17		100+	-
	South African National Roads Agency Ltd (SANRAL)	Royal HaskoningDHV		eThekwini Transport Authority	-

(Continued)

Name	Rea Vaya Phase 1A	Rea Vaya Phase 1B	Rea Vaya Phase 1C	MyCiTi Phase 1A	MyCiTi Phase 1B	MyCiTi Phase 2
Impact assessment	85% of Johannesburg's inhabitants will find a Rea Vaya Trunk or Feeder service within 500 meters of their place of residence or work			to provide services for 75% of households within a 500 meter radius		
Bus Operating Company	PioTrans (Pty): 100% owned by 313 shareholders; formed from 18 taxi associations, in February 2011; 585 taxis withdrawn	Litsamaiso: formed from 10 taxi associations (74%), Putco (22%), Metrobus (4%), in April 2014, 317 taxis withdrawn	associations have yet to be identified	TransPeninsula Investments (TPI) operates along the Waterfront; Kidrogen Vehicle Operating Company (VOC) operates between Dunoon and the Table View, Parklands and Melkbosstrand suburbs (removed 62 minibus taxis); Golden Arrow operates to Table View	Golden Arrow Bus Services, Route 6 Taxi Association for Mitchells Plain and Codeta, operating from Khayelitsha	

A Re Yeng Inception Phase	A Re Yeng (entire network)	Yarona Phase 1	Yarona (entire network)	Go Durban! Phase 1	Libhongolethu
85% of Tshwane's population will be within 500 metres of the BRT trunk or feeder		BRT trunk or feeder will reach at least 85% of Rustenburg residents within 1 kilometer of their homes.		85% of all residents within will reside within a 1 kilometer radius of a station by 2020 (move 25% of the Municipality's total trunk public transport demand on road-based IRPTN services)	"To provide an efficient, safe, affordable, sustainable and accessible multi-modal public transport system which supports social and economic development to ensure optimal mobility and improved quality of life for the residents and users of the transport system in the metropolitan area".
Tshwane Rapid Transit Pty, Ltd. (formed September 2012)		Rustenburg Taxi Industry (9 associations for phase 1; 21 taxi associations - 11 participating taxi associations; bus companies - Bojanala Bus and Thari Bus)	assocations have yet to be identified	Memorandum of Agreement signed 14 February 2014	TransBay signed 12 month contract in 2013; five routes would be operated by five companies, each comprising bus and taxi operators

(Continued)

Name	Rea Vaya Phase 1A	Rea Vaya Phase 1B	Rea Vaya Phase 1C	MyCiTi Phase 1A	MyCiTi Phase 1B	MyCiTi Phase 2
Automatic Fare Card (AFC) launch date	July 2013 (in partnership with ABSA); cards given free until February 2014			February 2012 in partnership with ABSA		
Raised lane delineator	rumble blocks - concrete	rumble blocks - polymer	rumble blocks - polymer	Kassel curb		
Asphalt	red painted line through the center of the roadway retrofitted in February 2012			red asphalt included in original design		
Number of buses	143 (41 18m- articulated, 102 13m-rigid, double door)	134 (41 18m-articulated; 93 13-m, double door)	240 (69 18m- articulated; 171 13m- standard buses)	267 (130 articulated; 87 12m-standard; 50 9m-buses)	-	-
Taxi Transition Consultants	Future of Transport (Darko Skibersek)	Future of Transport (Darko Skibersek)	-	David Schmidt for TPI	-	-
Bus ownership	Operator owns buses	City owns buses for first 5 years & then buses transferred to B.O.C. at market value	-	The vehicles purchased by the City out of the grant will be held as City assets and made available to the VOCs at no cost, to be used and maintained by the VOCs for the length of the contract, apart from the initial period, during which the vehicle supplier may be contracted to maintain the vehicles.		

APPENDIX B: FEATURES OF BRT SYSTEMS IN SOUAH AFRICA   161

A Re Yeng Inception Phase	A Re Yeng (entire network)	Yarona Phase 1	Yarona (entire network)	Go Durban! Phase 1	Libhongolethu
Jul-14		out for tender in July 2014		Muvo smart card launched May 2012 in partnership with Standard Bank	-
Kassel curb		raised roadway		curb	curb
red painted line through the center of the roadway		No red line		red asphalt included in original design	red painted line through the center of the roadway
5x 12m standard; 39x 18m articulated diesel powered Euro V buses	171 buses (46 articulated; 125 rigid)	260 buses (34 22marticulated; 226 13.9mstandard)	854 (99 22marticulated; 640 13.9mstandard; 115 midi-buses)	189 articulated; 226 standard; 562 midi (124 buses purchased from Dec '11-Apr'12 for R240 million of which 34 are BRT buses)	On 29 January 2013 NMBM handed over 19 buses, bought during the 2010 Soccer World Cup, to a consortium as the start of the IPTS
TBD		Future of Transport (Darko Skibersek)		team appointed in August 2013	Paul Browning
B.O.C. purchased buses		Intention for B.O.C. to own the buses		City owns buses	City owns buses

(Continued)

Name	Rea Vaya Phase 1A	Rea Vaya Phase 1B	Rea Vaya Phase 1C	MyCiTi Phase 1A	MyCiTi Phase 1B	MyCiTi Phase 2
Number of seats on the bus	117 passengers - 61 standing, 56 seated			articulated - 53 seating, 77 standing; 12m - 43 seating, 44 standing, 9m - 25 seated, 25 standing	-	-
Height of the stations (from road surface)	high-floor - 940mm		mixture of high-floor and low-floor	high-floor		low-floor
Number of stations	30	15	13	17	-	-
Station theme	iconic with local art			sustainable		
Station architect	Ikemeleng Architects - Jonathan Manning				-	-
Depot	Meadowlands Bus Depot, Dobsonville (8 hectares)	Selby (5 hectares)	Ivory Park or Greenstone (5 hectares)	3 depots	-	-
Trunk length (in km)	25.5	18	16	16	-	-
Total length including feeder services (in km)	122	93.22 km in inclusive trunk, feeder and complementary routes, covering 193.11 km in total	-	-	-	-

APPENDIX B: FEATURES OF BRT SYSTEMS IN SOUAH AFRICA   163

A Re Yeng Inception Phase	A Re Yeng (entire network)	Yarona Phase1	Yarona (entire network)	Go Durban! Phase 1	Libhongolethu
18m carry 90 passengers with 44 seats; 12m carry 60 passengers	-	-		articulated carry 160; standard carry 60 passengers; 30/16 seats on midi	150 articulated 18 m low-floor buses (total of 400 buses)
low-floor - 340 mm		low floor		low-floor	low-floor
7	60	13 BRT stations 3 CBD stations, 10 other stations); 268 bus stops	31 stations + central station	42 stations + 297 bus stops	-
Retro tram concept		mining with green inspiration		futuristic	-
PKA International Architects/Makeka Design Lab		ARG Architects		Iyer Urban Design Studio	-
3		2 (Tlhabane; Boitekong)	4 (Phokeng, Kanana, Tlhabane, Boitekekong)	4 terminal stations + 16 depots	-
5	70	Corridor A is 5 km; Corridor B is 7.5 km	-	56.5	10
190 (70 trunk + 120 feeder)		-	-	690 (57 trunk + 181 feeder + 452 complementary)	-

(Continued)

164 APPENDIX B: FEATURES OF BRT SYSTEMS IN SOUAH AFRICA

Name	Rea Vaya Phase 1A	Rea Vaya Phase 1B	Rea Vaya Phase 1C	MyCiTi Phase 1A	MyCiTi Phase 1B	MyCiTi Phase 2
Cost per kilometer	R36 million	R35 million	R35 million (not including stations, bridges & NMT)	R40 million	R4.5 billion total capital expenditure; R400 million operating deficit per annum	-
Cost per station	R12 million per module	R12 million per module	-	R4 million per station	-	-
Cost per trip	R25 (smart card); R11.50 (25km-35km); R12.50 (35+ km); R13 (occasional user one- trip); R25 (occasional user two-trip)			R25 (myconnect card); R6.80 (0-5km); R7.90 (5-10km); R9.60 (1020km); R12.70 (2030km); R14.30 (30-60km); R21.10 (>60km)		
Weekday patronage	43,000 (over 1 million per month) (5,760 passengers per direction per hour in peak)	-	-	30,000	-	-
Anticipated ridership	80,000	65,000	200,000 forecasted total for Phase 1	100,000 expected patronage	89 % increase from 400,000 total riders in November 2013 to 761,000 in February 2014	-

APPENDIX B: FEATURES OF BRT SYSTEMS IN SOUAH AFRICA   165

A Re Yeng Inception Phase	A Re Yeng (entire network)	Yarona Phase1	Yarona (entire network)	Go Durban! Phase 1	Libhongolethu
R50 million		R 680 million for bus fleet & R2.337 bilion for infrastructure	R2 billion for buses & R5 billion for infrastructure	R58.5m on upgrading its four bus depots in Umlazi, Ntuzuma, Rossburgh and KwaMashu;	R1.3 billion allocated before 2010
R10 million per station		R8 million per station		-	-
zonal system with commuters paying a flat rate of R10 for each initial boarding (trunk, complimentary and feeders) and a distance-based kilometre trip of R0.20/km	-		-		Fares before 13h00 will vary from R8 and R11 depending on distance; after 13h00 the fares are reduced to R7.50, R9.40 and R10.30 depending on distance.
-	-	-	-	-	-
10,000	127,000	225,000 daily passenger trips (45% of full network)	500,000	C1: 212,856 trips; C3: 129,409 trips; C9: 74,693 trips	-

(Continued)

References

ALC-BRT and Embarq. 2013. Global BRT data. Available at: brtdata.org (Accessed: October 28, 2016).

Allen, J. 2003. *Lost Geographies of Power*. Blackwell Publishers Ltd, Malden, Massachusetts.

Anderson, J.E. 1996. *Some Lessons from the History of Personal Rapid Transit (PRT)*. Washington: University of Washington, 1–24. Available at: http://faculty.washington.edu/jbs/itrans/history.htm (Accessed: 20 September 2013).

Ardila, A. 2007. How public transportation's past is haunting its future in Bogota, Colombia. *Journal of the Transportation Research Board* 2038, 9–15.

Ardila-Gomez, A. 2004. *Transit Planning in Curitiba and Bogotá: Roles in Interaction, Risk and Change*. Doctoral Dissertation. Massachusetts Institute of Technology.

Asmal, Z. (ed.). 2012. *Reflections and Opportunities: Design, Cities and the World Cup*. Designing South Africa, Cape Town.

Beall, J., Crankshaw, O., and Parnell, S. 2002. *Uniting a Divided City: Governance and Social Exclusion in Johannesburg*. Earthscan, London.

Beavon, K. 2004. *Johannesburg: The Making and Shaping of the City*. UNISA Press, Pretoria.

de Beer, L. 2004a. Detailed planning of the Strategic Public Transport Network (SPTN) Flagship Component: Soweto-Parktown – Motivation for future funding. Johannesburg, 3 March.

de Beer, L. 2004b. Detailed planning of the Strategic Public Transport Network (SPTN) Flagship Component: Soweto-Parktown – Working Group Meeting 3. Johannesburg, 27 May.

de Beer, L. 2005. *Strategic Public Transport Network: Update*. Johannesburg, 28 September.

Behrens, R. 2014. Urban mobilities: innovation and diffusion of public transport, in Parnell, S., and Oldfield, S. (eds.), *The Routledge Handbook on Cities of the Global South*, 459–473, 20–14, Routledge, London.

Behrens, R., and Del Mistro, R. 2010. Shocking habits: Methodological issues in analyzing changing personal travel behavior over time. *International Journal of Sustainable Transport* 4, 5, 253–271.

Behrens, R., McCormick, D., and Mfinanga, D. 2015. *Paratransit in African Cities: Operations, regulation and reform*. Earthscan, New York, 2015.

Behrens, R., and Wilkinson, P. 2003. Housing and urban passenger transport policy and planning in South Africa cities: A problematic Relationship, in Harrison, P., Huchzermeyer, M., and Mayekiso, M. (eds.), *Confronting Fragmentation. Housing and Urban Development in a Democratising Society*. UCT Press, Cape Town, 155–174.

Beltran, P., Gschwender, A., and Palma, C. 2013. The impact of compliance measures on the operation of a bus system: The case of Transantiago. *Research in Transportation Economics* 39, 79–89.

Bénit-Gbaffou, C., Didier, S., and Peyroux, E. 2012. Circulation of security models in Southern African Cities: Between Neoliberal Encroachment and Local Power Dynamics: Security models in Southern African cities. *International Journal of Urban and Regional Research*, 36, 5, 877–889. doi:10.1111/j.1468-2427.2012.01134.x.

Bennett, C.J. 1991. "What is Policy Convergence and What Causes it? *British Journal of Political Science* 21, 215–233.

Bickford, G., and Behrens, R. 2015. What does Transit Oriented Development mean in a South African Context? A Multiple Stakeholder Perspective from Johannesburg, in *Proceedings of the 34th Southern African Transport Conference*. Tshwane, South Africa, pp. 1–13.

Bray, D.J., Taylor, M.A.P., and Scrafton, D. 2011. Transport policy in Australia – Evolution, Learning and policy transfer. *Transport Policy* 18, 522–532.

Brenner, N., Peck, J., and Theodore, N. 2010. Variegated Neoliberalization: Geographies, modalities, pathways. *Global Networks* 10, 2, 182–222.

Brenner, N., and Theodore, N. 2002. Cities and the Geographies of "Actually Existing Neoliberalism". *Antipode* 34, 3, 349–379.

Browning, P. 2001. Wealth on Wheels? The Minibus-Taxi, Economic Empowerment and the New Passenger Transport Policy, in. *Southern Africa Transport Conference*, Tshwane, South Africa: CSIR, pp. 1–10.

Browning, P. 2005. The Minibus-Taxi in planned and integrated transport systems, in *Integrated Transport Planning*. Johannesburg, 1–9.

Browning, P. 2006. The paradox of the Minibus-Taxi: Towards a resolution of a paradox, in *South Africa Road Federation Transport Seminar*. Tshwane, 9 November.

Bunnell, T., and Das, D. 2010. A geography of serial seduction: Urban policy transfer from Kuala Lumpur to Hyderabad. *Urban Geography* 31, 3, 277–284.

Burawoy, M. 1998. The extended case method. *Sociological Theory* 16, 4, 4–33.

Burawoy, M. 2009. *The Extended Case Method: Four Countries, Four Decades, Four Great Transformations and One Theoretical Tradition*. University of California Press, Berkeley.

Büscher, M., and Urry, J. 2009. Mobile methods and the empirical. *European Journal of Social Theory* 12, 1, 99–116.

Cain, A., Darido, G., Baltes, M.R., Rodriguez, P., and Barrios, J.C. 2007. Applicability of TransMilenio Bus Rapid Transit System of Bogotá, Colombia, to the United States. *Journal of the Transportation Research Board* 2034, 45–54.

Callaghan, L., and Vincent, W. 2007. Preliminary evaluation of Metro Orange Line Bus Rapid Transit Project. *Journal of the Transportation Research Board* 2034, 37–44.

Cameron, B. 2007. The Time Has Come…To Talk of Many Things…Of Empires and Silos…And Study Tour Flings, in *BRT Workshop. Southern Africa Transport Conference*, Tshwane, South Africa.

Campbell, T. 2012. *Beyond Smart Cities: How Cities Network, Learn and Innovate.* Earthscan, New York.

Carpiano, R.M. 2009. Come Take a Walk with Me: The "Go-Along" Interview as a Novel Method for Studying the Implications of Place for Health and Well-Being. *Health & Place* 15, 263–272.

Castells, M. 1996. *The Rise of the Network Society.* Blackwell Publishers Ltd, Oxford.

Cervero, R., and Kang, C.D. 2011. Bus Rapid Transit Impacts on Land Use and Land Values in Seoul, Korea. *Transport Policy* 18, 102–116.

Cherry, G.E. (ed.). 1981. *Pioneers in British Planning.* The Architectural Press, London.

Chipkin, C. 1993. *Johannesburg Style: Architecture and Society 1880s–1960s.* David Philip Publishers, London.

Choy, T.K., Faier, L., Hathaway, M., Inoue, M., Satsuka, S., and Tsing, A. 2009. A New Form of Collaboration in Cultural Anthropology: Matsutake Worlds. *American Ethnologist* 36, 2, 380–403.

Christopher, A.J. 1995. Segregation and Cemeteries in Port Elizabeth, South Africa. *The Geographical Journal* 161, 1, 38–46.

Chua, B.H. 2011. Singapore as Model: Planning innovations, knowledge experts, in Roy, A. and Ong, A. (eds.), *World Cities: Asian Experiments and the Art of Being Global.* Wiley-Blackwell, Malden, Massachusetts, 29–54.

City of Cape Town. 2003. *Provincial Vision for Public Transport and Five-Year Strategic Delivery Programme.* Provincial Administration of the Western Cape, Cape Town.

City of Cape Town. 2010. *Phase 1A of Cape Town's MyCiTi Integrated Rapid Transit System: Business Plan.* Department of Transport, Cape Town.

City of Cape Town. 2016. *Transit Oriented Development Strategic Framework.* Transport and Urban Development Authority, Cape Town.

City of Cape Town Department of Transport, Roads and Stormwater 2012. Transport, Roads and Stormwater's Transformation and Action Plan 2012.

City of Cape Town MyCiTi Project Office. 2012. *2012 MyCiTi Business Plan: Phases 1A, 1B and N2 Express of Cape Town's MyCiTi IRT System.* City of Cape Town, Cape Town.

City of Johannesburg. 2011. *Rea Vaya/BRT – Transforming the Face of Public Transport: End of Term Report 2006–2011.* Department of Transport, Johannesburg.

City of Johannesburg Transportation Department. 2011. *Rea Vaya/BRT – Transforming the Face of Public Transport. End of Term Report: 2006 – 2011.* Johannesburg Transportation Department, Johannesburg.

Clark, P., and Crous, W. 2002. Public transport in Metropolitan Cape Town: Past, present and future. *Transport Review: A Transnational Transdisciplinary Journal* 22, 1, 77–101.

Clarke, N. 2009. In What Sense "Spaces of Neoliberalism"? The new Localism, the New Politics of Scale, and Town Twinning. *Political Geography* 28, 496–507.

Clarke, N. 2010. Town Twinning in Cold-War Britain: (Dis)continuities in Twentieth-century municipal internationalism. *Contemporary British History* 24, 2, 173–191.

Clarke, N. 2012. Urban policy mobility, Anti-politics, and histories of the transnational municipal movement. *Progress in Human Geography* 36, 1, 25–43.

Cook, I.M., and Harrison, M. 2007. Follow the Thing: "'West Indian Hot Pepper Sauce'". *Space and Culture* 10, 1, 40–63.

Cook, I.R. 2008. Mobilising urban policies: The Policy Transfer of U.S. Business improvement districts to England and Wales. *Urban Studies* 45, 4, 773–795.

Cook, I.R., and Ward, K. 2011. Trans-urban networks of learning, mega events and policy tourism: The case of Manchester's Commonwealth and Olympic Games Projects. *Urban Studies* 48, 12, 2519–2535.

Cook, I.R., and Ward, K. 2012. Conferences, informational infrastructures and mobile policies: The process of getting Sweden "BID Ready". *European Urban and Regional Studies* 19, 2, 137–152.

Cook, I.R., Ward, S.V., and Ward, K. 2014a. A springtime Journey to the Soviet Union: Postwar planning and policy mobilities through the Iron Curtain. *International Journal of Urban and Regional Research* 38, 3, 805–822.

Cook, I.R., Ward, S.V., and Ward, K. 2014b. A Springtime Journey to the Soviet Union: Post-war Planning and policy mobilities through the Iron Curtain. *International Journal of Urban and Regional Research* 38, 3, 805–822.

Cooke, S., Behrens, R., and Zuidgeest, M. 2018. The relationship between transit-oriented development, accessibility and public transport viability in South African cities: A literature review and problem framing, in *Proceedings of the 37th Southern African Transport Conference*. Tshwane, South Africa, pp. 365–380.

Cresswell, T. 2006. *On the Move: Mobility in the Modern Western World*. Routledge, New York.

Crot, L. 2010. Transnational urban policies: "Relocating" Spanish and Brazilian models of urban planning in Buenos Aires. *Urban Research & Practice* 3, 2, 119–137.

Davies, R.J. 1981. The spatial formation of the South African City. *GeoJournal* 2, 59–72.

Deleuze, G., and Guattari, F. 1972. *Anti-Oedipus: Capitalism and Schizophrenia*. Translated by Hurley, R., Seem, M., and Lane, H. Continuum, London.

Deleuze, G., and Guattari, F. 2004. *A Thousand Plateaus: Capitalism and Schizophrenia*. Continuum, London.

Delmelle, E.C., and Casas, I. 2012. Evaluating the Spatial Equity of Bus Rapid Transit-based Accessibility Patterns in a Developing Country: The Case of Cali, Colombia. *Transport Policy* 20, 36–46.

Deng, T., and Nelson, J.D. 2011. Recent developments in Bus Rapid Transit: A Review of the Literature. *Transport Reviews* 31, 1, 69–96.

Deng, T., and Nelson, J.D. 2013. Bus Rapid Transit implementation in Beijing: An evaluation of performance and impacts. *Research in Transportation Economics* 39, 108–113.

Dentlinger, L. 2016. Cape Town transport authority reinvented. *Cape Argus*, 24 October. Available at: https://www.iol.co.za/motoring/industry-news/cape-town-transport-authority-reinvented-2082904 (Accessed: 19 May 2020).

Department of Transport. 1999. *Moving South Africa: A Transport Strategy for 2020*. Department of Transport, Pretoria.

Department of Transport. 2006. *Towards A 2010 Transport Action Agenda: A Discussion Document to Enable Transportation Success for the 2010 FIFA World Cup and to Create A Lasting Legacy for Public Transport*. Department of Transport, Tshwane, South Africa.

Department of Transport. 2007. *Public Transport Action Plan*. Department of Transport, Pretoria.

Didier, S., Peyroux, E., and Morange, M. 2012. The Spreading of the City Improvement District Model in Johannesburg and Cape Town: Urban Regeneration and the Neoliberal Agenda in South Africa. *International Journal of Urban and Regional Research* 36, 5, 915–935.

Dogliani, P. 2002. European Municipalism in the First Half of the Twentieth Century: The Socialist Network. *Contemporary European History* 11, 4, 573–596.

Dolowitz, D., and Marsh, D. 1996. Who Learns What from Whom: A Review of the Policy Transfer Literature. *Political Studies* 44, 343–357.

Dolowitz, D., and Marsh, D. 2000. Learning from Abroad: The Role of Policy Transfer in Contemporary Policy-Making. *Governance: An International Journal of Policy and Administration* 13, 1, 5–24.

Dreyfus, H.L., and Rabinow, P. (eds.). 1982. *Michael Foucault: Beyond Structuralism and Hermeneutics*. Harvester, Brighton.

Dugard, J. 2001. From Low Intensity War to Mafia War: Taxi Violence in South Africa (1987–2000). *Violence and Transition* 4.

Evans, J., and Jones, P. 2011. The Walking Interview: Methodology, Mobility and Place. *Applied Geography* 31, 849–858.

Farias, I., and Bender, T. (eds.). 2010. *Urban Assemblages: How Actor-Network Theory Changes Urban Studies*. Routledge, London.

Ferbrache, F. (ed.). 2019. *Developing Bus Rapid Transit: The Value of BRT in Urban Spaces*. Edward Elgar Publishing, Northampton, Massachusetts.

Filipe, L.N., and Macario, R. 2013. A First Glimpse on policy packaging for implementation of BRT Projects. *Research in Transportation Economics* 39, 150–157.

Florida, R. 2002. *The Rise of the Creative Class: And How It's Transforming Work*. Routledge, London.

Foucault, M. 1980. *Power/Knowledge: Selected Interviews and Other Writings, 1972–1977*. Translated by Gordon, C. New York: Pantheon Books.

Foucault, M. 2003. *The Essential Foucault: Selections from Essential Works of Foucault 1954–1984*. Edited by Rabinow, P. and Rose, N. The New Press, London.

Froschauer, P. 2012. Rustenburg rapid Transit – Taking transport further. April.

Gaspari, O. 2002. Cities against states? Hopes, Dreams and Shortcomings of the European Municipal Movement, 1900–1960. *Contemporary European History* 11, 597–621.

Gauteng Department of Roads and Transport. 2020. *Gauteng Province Household Travel Survey Report*. Gauteng Department of Roads and Transport, Tshwane, South Africa.

Gilbert, A. 2008. Bus Rapid Transit: Is Transmilenio a Miracle Cure? *Transport Reviews* 28, 4, 439–467.

Gilbert, A., and Crankshaw, O. 1999. Comparing South African and Latin American Experience: Migration and Housing Mobility in Soweto. *Urban Studies* 36, 13, 2375–2400.

Gill, F. 1961. *Cape Trams: From Horse to Diesel*. Cape Electric Tramways, Cape Town.

Goldman, M. 2005. *Imperial Nature*. Yale University Press, New Haven.

González, S. 2011. Bilbao and Barcelona "In Motion": How Urban Regeneration "Models" Travel and Mutate in the Global Flows of Policy Tourism. *Urban Studies* 48, 7, 1397–1418.

Haferburg, C., and Huchzermeyer, M. (eds.). 2014. *Urban Governance in Post-Apartheid Cities: Modes of Engagement in South Africa's Metropoles*. Borntraeger Science Publishers, Stuttgart, Germany.

Hains, C. 2011. Cracks in the Facade: Landscapes of Hope and Desire in Dubai, in Roy, A. and Ong, A. (eds.), *World Cities: Asian Experiments and the Art of Being Global*. Wiley-Blackwell, Malden, Massachusetts, 160–181.

Hannam, K., Sheller, M., and Urry, J. 2006. Mobilities, Immobilities and Moorings. *Mobilities* 1, 1, 1–22.

Hansen, T.B. 2006. Sounds of Freedom: Music, taxis, and racial imagination in Urban South Africa. *Public Culture* 18, 1, 185–208.

Hansen, T.B. 2012. *Melancholia of Freedom: Social Life in an Indian Township in South Africa*. Princeton University Press, New Jersey.

Harradine, M. 1997. *Port Elizabeth: A Social Chronicle to the End of 1945*. E.H. Walton Packaging, Port Elizabeth.

Harrison, P., Gotz, G., Todes, A., and Wray, C. (eds.) 2014. *Changing Space, Changing City: Johannesburg after Apartheid*. Wits University Press, Johannesburg. doi: 10.18772/22014107656.

Harrison, P., Todes, A., and Watson, V. 2008. *Planning and Transformation: Learning from the Post-Apartheid Experience*. Routledge, London.

Hart, G.H.T. 1984. Urban Transport, Urban Form and Discrimination in Johannesburg. *South African Geographical Journal* 66, 2, 152–167.

Healey, P. 2013. Circuits of Knowledge and Techniques: The transnational flow of planning ideas and practices. *International Journal of Urban and Regional Research* 37, 5, 1510–1526.

Healey, P., and Upton, R. (eds.). 2010. *Crossing Borders: International Exchanges and Planning Practices*. Routledge, New York.

HHO Africa. 2000. *Koeberg Road Management Strategy: Traffic and Transportation Report*. 5866. Blaauwberg Municipality, Cape Town.

HHO Africa. 2003. *Conceptual Planning of the N1 Corridor between Monte Vista and Cape Town: Interim Report*. 6317. City of Cape Town Provincial Administration Western Cape, Cape Town.

HHO Africa. 2004. *Public Transport Planning Design and Implementation: Project 5.7: Public Transport Corridor Strategy: Summary Report*. 6532. City of Cape Town: Transport, Roads and Stormwater, Cape Town: City of Cape Town.

HHO Africa. 2007a. *Cape Town IRT Implementation Framework*. City of Cape Town Department of Transport, Roads and Stormwater, Cape Town.

HHO Africa. 2007b. *Conceptual Planning of the N1 Corridor between Bellville and Cape Town: Phase 2*. 6559. City of Cape Town Provincial Administration Western Cape, Cape Town.

Hidalgo, D., and Graftieaux, P. 2008. Bus Rapid Transit Systems in Latin America and Asia: Results and Difficulties in 11 Cities. *Journal of the Transportation Research Board* 2072, 77–88.

Hidalgo, D., and Gutiérrez, L. 2013. BRT and BHLS around the World: Explosive growth, large positive impacts and many issues outstanding. *Transport Economics* 39, 8–13.

Hodgson, P., Potter, S., Warren, J., and Gillingwater, D. 2013. Can bus really be the new tram? *Research in Transportation Economics* 39, 158–166.

Hoffman, L. 2011. Urban Modeling and contemporary technologies of City-Building in China: The production of regimes of Green Urbanisms, in Roy, A. and Ong, A. (eds.), *World Cities: Asian Experiments and the Art of Being Global*. Wiley-Blackwell, Malden, Massachusetts, 55–76.

Home, R.K. 1990. Town Planning and Garden Cities in the British Colonial Empire 1910–1940. *Planning Perspectives* 5, 23–37.

Hook, W. 2014. Lessons and new directions as Bus Rapid Transit Turns 40. Available at: http://www.citiscope.org/story/2014/lessons-and-new-directions-bus-rapid-transit-turns-40 (Accessed: 10 April 2014).

Hoyt, L. 2006. Importing Ideas: The transnational transfer of urban revitalization policy. *International Journal of Public Administration* 29, 221–243.

IOL News. 2010. Controversial Taxi Boss Shot Dead in Tavern. 15 May.

ITDP. 2013. History of ITDP. 25 January. Available at: www.itdp.org/who-we-are/history-of-itdp. August 15, 2016.

ITDP. 2014. Access for All: TransJakarta Improvements Maximize the Benefits of BRT. Available at https://www.itdp.org/2014/02/10/access-for-all-transjakarta-improvementsmaximize-the-benefits-of-brt/ (Accessed: 10 April 2014).

ITDP. 2017. *The TOD Standard 3.0*. Institute for Transportation and Development Policy, New York.

ITDP. 2019. Transjakarta: A Study in Success. Available at https://www.itdp.org/2019/07/15/transjakarta-study-success/ (Accessed: 10 October 2019).

Jackson, A. 2003. *Public Transport in Durban – A Brief History*. Available at: http://www.fad.co.za/Resources/transport/transport.htm. January 15, 2016.

Jackson, T. 2016. Can big data offer hope for Africa's exasperated commuters? *Guardian* 7 October.

Jarzab, J.T., Lightbody, J., and Maeda, E. 2002. Characteristics of Bus rapid transit projects: An overview. *Journal of Public Transportation* 5, 2, 31–46.

Jennings, G. 2015. A bicycling renaissance in South Africa? Policies, programmes and trends in Cape Town, in *Proceedings of the 34th Southern African Transport Conference.* Tshwane, South Africa, pp. 486–498.

Jessop, B., and Peck, J. 1998. Fast Policy/Local Discipline: The Politics of Time and Scale in the Neoliberal Workfare Offensive. *Mimeograph Department of Sociology*, Lancaster University.

Joyce, P. (ed.). 1981. *South Africa's Yesterdays.* Reader's Digest Association of South Africa`, Johannesburg.

Kane, L. 2003. Mayor of Bogota's Bold Vision for Cape Town that would favour the Pedestrian Over the Freeway. *Cape Times* 31 January.

Kane, L., and Del Mistro, R. 2003. Changes in transport planning policy: Changes in transport planning methodology. *Transportation* 30, 2, 113–131.

Khosa, M. 1991. Routes, ranks and rebels: Feuding in the Taxi revolution. *Journal of Southern African Studies* 18, 1.

Khosa, M. 1995. Transport and popular struggles in South Africa. *Antipode* 27, 2, 167–188.

King, A.D. 1976. *Colonial Urban Development: Culture, Social Power and Environment.* Routledge & Kegal Paul, London.

Kingdon, J.W. 1995. *Agendas, Alternatives and Public Policies.* Addison-Wesley Educational Publishers, Inc, Ann Arbor, Michigan.

Kozinska-Witt, H. 2002. The Union of Polish Cities in the Second Polish Republic, 1918–1939: Discourses of Local Government in a Divided Land. *Contemporary European History* 11, 4, 549–571.

Kumar, A., Zimmerman, S., and Agarwal, O.P. 2012. *International Experience in Bus Rapid Transit (BRT) Implementation: Synthesis of Lessons Learned from Lagos, Johannesburg, Jakarta, Delhi, and Ahmedabad Case Studies.* The World Bank Institute, Washington, D.C., 57.

Larner, W. 2002. Globalization, Governmentality and Expertise: Creating a call centre labour force. *Review of International Political Economy* 9, 4, 650–674.

Larner, W., and Laurie, N. 2010. Travelling Technocrats, Embodied Knowledges: Globalising Privatisation in Telecoms and Water. *Geoforum* 41, 2, 218–226.

Latour, B. 1993. *We Have Never Been Modern.* Translated by Porter, C. Harvester Wheatsheaf, Hemel Hempstead.

Latour, B. 1996. *Aramis, or, the Love of Technology.* Translated by Porter, C. Cambridge, Massachusetts: Harvard University Press.

Latour, B. 2005. *Reassembling the Social: An Introduction to Actor-Network-Theory.* Oxford University Press, Oxford.

Laurier, E. 2004. Doing office work on the motorway. *Theory, Culture & Society* 21, 261–277.

Laurier, E. 2008. Driving and "passengering": Notes on the ordinary organization of car travel. *Mobilities* 3, 1–23.

Law, J., and Urry, J. 2004. Enacting the Social. *Economy and Society* 33, 3, 390–410.

Lefebvre, H. 1991. *The Production of Space*. Translated by Nicholson-Smith, D. Oxford: Blackwell Publishers Ltd.

Lemon, A. (ed.). 1991. *Homes Apart: South Africa's Segregated Cities*. Indiana University Press, Bloomington.

Lester, A. 2006. Imperial circuits and networks: Geographies of the British Empire. *History Compass* 4, 1, 124–141.

Levinson, H.S., Zimmerman, S., Clinger, J., and Gast, J. 2003. Bus Rapid Transit: Synthesis of case studies. *Journal of the Transportation Research Board* 1841, 1–11.

Litman, T. 2006. Innovative Transport Solutions. *Southern Africa Transport Conference*, Tshwane, 12 July.

Mahadevia, D., Joshi, R., and Datey, A. 2013. Ahmedabad's BRT System: A sustainable urban transport Panacea? *Review of Urban Affairs* 48, 56–64.

Mahon, R., and Macdonald, L. 2010. Anti-poverty politics in Toronto and Mexico City. *Geoforum* 41, 2, 209–217.

Manning, J. 2007. *Bus Rapid Transit – Workshop on Station Design*. Ikemeleng Architects, Johannesburg.

Manning, J. 2009. *Rea Vaya BRT Phase 1 Planning Guidelines*. Ikemeleng Architects, Johannesburg.

Marcus, G.E. 1995. Ethnography In/Of the World System: The emergence of Multi-sited ethnography. *Annual Review of Anthropology* 24, 95–117.

Marsden, G., Frick, K.T., May, A.D., and Deakin, E. 2011. How do Cities Approach Policy Innovation and Policy Learning? A Study of 30 Policies in Northern Europe and North America. *Transport Policy* 18, 501–512.

Marsden, G., and Stead, D. 2011. Policy transfer and learning in the field of transport: A review of concepts and evidence. *Transport Policy* 18, 492–500.

Masondo, A. 2008. The official opening of the BRT prototype station. Johannesburg, May. Available at: http://www.joburg.org.za/index.php?option=com_content&id=3154&.#ixzz33lLPhWDy. June 5, 2015.

Massey, D. 2011. A Counterhegemonic Relationality of Place, in McCann, E. and Ward, K. (eds.), *Mobile Urbanism: Cities and Policymaking in the Global Age*. University of Minnesota Press, Minneapolis, 1–14.

Matsumoto, N. 2007. Analysis of Policy Processes to Introduce Bus Rapid Transit Systems in Asian Cities from the Perspective of Lesson-drawing: Cases of Jakarta, Seoul, and Beijing, in Institute for Global Environmental Strategies (ed.), *Air Pollution Control in the Transportation Sector: Third Phase Research Report of the Urban Environmental Management Project*. Institute for Global Environmental Strategies, Japan, 359–376.

Matsumoto, N., King, P., and Mori, H. 2007. Policies for environmentally sustainable transport. *International Review for Environmental Strategies: Best Practice on Environmental Policy in Asia and the Pacific* 7, 1, 97–116.

Maunganidze, L., and Del Mistro, R. 2012. The Role of Bus Rapid Transit in Improving Public Transport Levels of Service, Particularly for the Urban Poor Users of

Public Transport: A Case of Cape Town, South Africa, in *Southern Africa Transport Conference*, Tshwane, South Africa.

May, A.D. 1995. The design of integrated transport strategies. *Transport Policy* 2, 1, 97–105.

Maylam, P. 1995. Explaining the Apartheid city: 20 Years of South African urban historiography. *Journal of Southern African Studies* 21, 1, 19–38.

Mazaza, M. 2017. *Comprehensive integrated transport plan 2018–2023*. City of Cape Town: Transport and Urban Development Authority. Available at: https://tdacontenthubstore.blob.core.windows.net/resources/fd3ddc0d-b459-4d26-bb01-7f689d7a36eb.pdf (Accessed: 19 May 2020).

Mazaza, M., and Petersen, R. 2003. *Integrated Planning Progress Report*. Department of Transport, Cape Town.

Mbembe, A. 2004. Aesthetics of Superfluidity. *Public Culture* 16, 373–405.

Mbembe, A. 2008. Aesthetics of Superfluity, in Nuttall, S. and Mbembe, A. (eds.), *Johannesburg: The Elusive Metropolis*. Duke University Press, London, 37–67.

Mbembe, A., and Nuttall, S. 2004. Writing the World from an African Metropolis. *Public Culture* 16, 3, 347–372.

McCann, E. 2004. "Best Places": Interurban competition, quality of life and population media discorse. *Urban Studies* 41, 1909–1929.

McCann, E. 2008. Expertise, truth, and urban policy mobilities: Global circuits of knowledge in the development of Vancouver, Canada's "Four Pillar" Drug strategy. *Environment and Planning A* 40, 885–904.

McCann, E. 2011. Urban policy mobilities and global circuits of knowledge: Toward a research Agenda. *Annals of Association of American Geographers* 101, 1, 107–130.

McCann, E. 2013. Policy boosterism, policy mobilities, and the Extrospective city. *Urban Geography* 34, 1, 5–29.

McCann, E., and Ward, K. 2010. Relationality/Territoriality: Towards a conceptualization of cities in the world. *Geoforum* 41, 2, 175–184.

McCann, E., and Ward, K. (eds.). 2011. *Mobile Urbanism: Cities and Policymaking in the Global Age*. University of Minnesota Press, Minneapolis.

McCann, E., and Ward, K. 2012. Assembling urbanism: Following policies and "Studying through" the sites and situations of policy making. *Environment and Planning A* 44, 1, 42–51.

McCann, E., and Ward, K. 2013. A Multi-disciplinary approach to policy transfer research: Geographies, assemblages, mobilities and mutations. *Policy Studies* 34, 1, 2–18.

McCaul, C. 1990. *No Easy Ride: The Rise and Future of the Black Taxi Industry*. South African Institute of Race Relations, Johannesburg.

McCaul, C. 2006a. *Johannesburg's Decision to Implement Bus Rapid Transit and the Vision*. Johannesburg, 30 August.

McCaul, C. 2006b. *The Introduction of the Rea Vaya BRT System in Johannesburg: Developments, Achievements and Challenges.* November, Johannesburg.

McCaul, C. 2008. *GIZ's Contribution to the Johannesburg Rea Vaya Bus Rapid Transit Project.* College McCaul Associates, Johannesburg.

McCaul, C. 2011. Negotiating the Deal to Enable the First Rea Vaya Bus Company: Agreements, Experiences and Lessons, in *Southern Africa Transport Conference*, Tshwane.

McFarlane, C. 2009. Translocal assemblages: Space, power and social movements. *Geoforum* 40, 4, 561–567.

McFarlane, C. 2011a. *Learning the City: Knowledge and Translocal Assemblage.* Wiley-Blackwell, Malden, Massachusetts.

McFarlane, C. 2011b. The city as a machine for learning: Translocalism, Assemblage and Knowledge. *Transactions of the Institute of British Geographers* 36, 3, 360–376.

Mejía-Dugand, S., Hjelm, O., Baas, L., and Rios, R.A. 2013. Lessons from the spread of bus rapid transit in Latin America. *Journal of Cleaner Production* 50, 82–90.

Mojica, C.H. 2011. TransMilenio: The battle over Avenida Séptima. *John F. Kennedy School of Government, Harvard University Case Program* 1942.0, 1–25.

Montero, S. 2016. Study tours and inter-city policy learning: Mobilizing Bogota's transportation policies in Guadalajara. *Environment and Planning A*. 49, 2, 332–350, 2016.

Montezuma, R. 2005. The Transformation of Bogota, Colombia, 1995–2000: Investing in Citizenship and Urban Mobility. *Global Urban Development* 1, 1, 1–10.

Moodley, N. 2007. Construction to start in September for R12bn Jo'burg-Soweto monorail, 16 May. Available at: https://www.engineeringnews.co.za/article/construction-to-start-in-september-for-r12bn-jo039burgsoweto-monorail-2007-05-16. (Accessed: 23 October 2015).

Morange, M., Folio, F. and Peyroux, E. 2012. The Spread of a Transnational Model: "Gated Communities" in Three Southern African Cities (Cape Town, Maputo and Windhoek). *International Journal of Urban and Regional Research* 36, 5, 890–914. doi:10.1111/j.1468-2427.2012.01135.x.

Mossberger, K., and Wolman, H. 2003. Policy transfer as a form of prospective policy evaluation: challenges and recommendations. *Public Administration Review* 63, 4, 428–440.

Myers, G. 2014. From expected to unexpected comparisons: Changing the flows of ideas about cities in a postcolonial urban world. *Singapore Journal of Tropical Geography* 35, 1, March, 104–118.

Nader, L. 1972. Up the Anthropologist-perspectives gained from studying up, in Hymes, D. (ed.), *Reinventing Anthropology.* Random House, New York, 284–311.

National Department of Transport. 1996. *White Paper on National Transport Policy.* Department of Transport, Pretoria.

National Department of Transport. 2000. *National Land Transport Transition Act.*

National Department of Transport. 2005. *National Household Travel Survey.* Department of Transport, Tshwane, South Africa.

National Department of Transport. 2009. *National Land Transport Act*.
National Treasury. 2003. *Intergovernmental Fiscal Review: Local Government Budgets and Expenditure 2001/2 to 2006/7*. National Treasury, Tshwane, South Africa.
National Treasury. 2011. *Intergovernmental Fiscal Review: Local Government Budgets and Expenditure 2006/7 to 2012/13*. National Treasury, Tshwane, South Africa.
National Treasury. 2012. *National Treasury Budget 2012: Estimates of National Expenditure*. National Treasury, Tshwane, South Africa, 1–80.
National Treasury. 2019. *2019 Budget Review*. National Treasury, Tshwane, South Africa.
Nelson Mandela Bay Steering Committee. 2010. *Nelson Mandela Bay Integrated Public Transport System*. Nelson Mandelay Bay.
Normack, D. 2006. Tending to Mobility: Intensities of staying at the petrol station. *Environment and Planning A* 38, 2, 241–252.
Norwich, O.I. 1986. *A Johannesburg Album: Historical Postcards*. AD Donker, Johannesburg.
Offe, C. 1996. Designing institutions in East European Transitions, in Goodin, R.E. (ed.), *The Theory of Institutional Design*. Cambridge University Press, Cambridge, UK, 199–226.
Olds, K. 1997. Globalizing Shanghai: The "Global Intelligence Corps" and the Building of Pudong. *Cities* 14, 2, 109–123.
Olds, K. 2001. *Globalization and Urban Change: Capital, Culture, and Pacific Rim Mega-Projects*. Oxford University Press, Oxford.
Ong, A. 2011. Introduction: World Cities, or the Art of Being Global, in Roy, A. and Ong, A. (eds.), *World Cities: Asian Experiments and the Art of Being Global*. Wiley-Blackwell, Malden, Massachusetts, 1–26.
Ong, A., and Collier, S.J. (eds.). 2005. *Global Assemblages: Technology, Politics, and Ethics as Anthropological Problems*. Blackwell Publishers Ltd, Malden, Massachusetts.
Osborne, T. 2004. On Mediators: Intellectuals and the Ideas Trade in Knowledge Society. *Economy and Society* 33, 4, 430–447.
Paget-Seekins, L., and Munoz, J.C. 2016. *Restructuring Public Transport through Bus Rapid Transit: An International and Interdisciplinary Perspective*. Policy Press, Bristol.
Parnell, S. 1997. South African Cities: Perspectives from the Ivory Tower of Urban Studies. *Urban Studies* 34, 5, 891–906.
Parnell, S., and Mabin, A. 1995. Rethinking Urban South Africa. *Journal of Southern African Studies* 21, 1, 39–61.
Parnell, S., and Pieterse, E. (eds.). 2014. *Africa's Urban Revolution*. Zed Books Ltd, Johannesburg.
Patton, B. 2002. *Double-Deck Trams of the World: Beyond the British Isles*. Adam Gordon, Brora, Sutherland.
Peck, J. 2002. Political economies of Scale: Fast policy, interscalar relations, and Neoliberal workfare. *Economic Geography* 78, 3, 331–360.
Peck, J. 2003. Geography and public policy: Mapping the Penal state. *Progress in Human Geography* 27, 2, 222–232.

Peck, J. 2011a. Creative moments: Working culture, through Municipal socialism and Neoliberal Urbanism, in McCann, E. and Ward, K. (eds.), *Mobile Urbanism: Cities and Policymaking in the Global Age*. University of Minnesota Press, Minneapolis, 41–70.

Peck, J. 2011b. Geographies of policy: from Transfer-Diffusion to Mobility-Mutation. *Progress in Human Geography* 36, 5, 1–25.

Peck, J., and Theodore, N. 2001. Exporting Workfare/Importing Welfare-to-Work: Exploring the politics of Third Way Policy transfer. *Political Geography* 20, 427–460.

Peck, J., and Theodore, N. 2010a. Mobilizing policy: Models, methods, and mutations. *Geoforum* 41, 2, 169–174.

Peck, J., and Theodore, N. 2010b. Recombinant workfare, across the Americas: Transnationalizing "Fast" social policy. *Geoforum* 41, 2, 195–208.

Peck, J., and Theodore, N. 2012. Follow the policy: A distended case approach. *Environment and Planning A* 44, 21–30.

Peck, J., and Theodore, N. 2015. *Fast Policy: Experimental Statecraft at the Thresholds of Neoliberalism*. University of Minnesota Press.

Peck, J., Theodore, N., and Brenner, N. 2009. Neoliberal urbanism: Models, moments, mutations. *SAIS Review of International Affairs* 29, 1, 49–66.

Penalosa, G. 2012. *Transforming Our Streets: Bringing the International Experience to Johannesburg*. University of Johannesburg, 16 May.

Peyroux, E., Pütz, R., and Glasze, G. 2012. Business Improvement Districts (BIDs): The internationalization and contextualization of a "travelling concept". *European Urban and Regional Studies* 19, 2, 111–120.

Pineda, A.V. 2010. How Do we Co-produce Urban transport systems and the City?: The case of TransMilenio and Bogota, in Farias, I. and Bender, T. (eds.), *Urban Assemblages: How Actor-Network Theory Changes Urban Studies*. Routledge, London, 123–138.

Pirie, G. 2013. Transport geography in South Africa. *Journal of Transport Geography* 31, 312–314.

Pirie, G. 2014. Transport pressures in urban Africa: Practices, policies, perspectives, in Parnell, S. and Pieterse, E. (eds.), *Africa's Urban Revolution*. Zed Books Ltd, London, 133–147.

Ponnaluri, R.V. 2011. Sustainable Bus Rapid Transit Initiatives in India: The role of decisive leadership and strong institutions. *Transport Policy* 18, 269–273.

Porto de Oliveira, O. 2017. *International Policy Diffusion and Participatory Budgeting: Ambassadors of Participation, International Institutions and Transnational Networks*. Palgrave Macmillan.

Pow, C.P. 2014. License to Travel. *City* 18, 3, 287–306.

Prince, R. 2012. Policy transfer, consultants and the Geographies of Governance. *Progress in Human Geography* 36, 2, 188–203.

Priya Uteng, T., and Lucas, K. (eds.). 2018. *Urban Mobilities in the Global South*. Routledge, Abingdon & New York.

Rabinovitch, J., and Hoehn, J. 1995. *A Sustainable Urban Transportation System: The 'Surface Metro' in Curitiba, Brazil.* Working Paper No. 19. United Nations Development Program, New York, 1–46.

Rabinow, P. (ed.). 1984. *The Foucault Reader.* Pantheon Books, New York.

Rapoport, E. 2015. Sustainable urbanism in the age of Photoshop: Images, experiences and the role of learning through inhabiting the international travels of a planning model. *Global Networks* 15, 3, 307–324.

Rapoport, E., and Hult, A. 2017. The travelling business of sustainable urbanism: International consultants as norm-setters. *Environment and Planning A* 49, 8, 1779–1796.

Robertson, D.B. 1991. Political Conflict and Lesson-Drawing. *Journal of Public Policy* 11, 1, 55–78.

Robins, S.L. 2008. *From Revolution to Rights in South Africa: Social Movements, NGOs, and Popular Politics after Apartheid.* James Currey, Suffolk, United Kingdom.

Robinson, J. 1996. *The Power of Apartheid: State, Power and Space in South African Cities.* Butterworth-Heinemann Ltd, Oxford.

Robinson, J. 1997. The geopolitics of South African cities: States, citizens, territory. *Political Geography* 16, 5, 365–386.

Robinson, J. 2003. Johannesburg's futures: Beyond developmentalism and global success, in Tomlinson, R., Beauregard, R., Bremmer, L., Mangcu, X. (ed.), *Emerging Johannesburg: Perspectives on the Postapartheid City.* Routledge, London.

Robinson, J. 2006. *Ordinary Cities: Between Modernity and Development.* Routledge, London.

Robinson, J. 2011. The Spaces of Circulating Knowledge: City Strategies and Global Urban Governmentality, in McCann, E. and Ward, K. (eds.), *Mobile Urbanism: Cities and Policymaking in the Global Age.* University of Minnesota Press, Minneapolis, 15–40.

Rose, N., and Miller, P. 1992. Political power beyond the state: Problematics of government. *The British Journal of Sociology* 43, 2, 173–205.

Rose, R. 1993. *Lesson Drawing in Public Policy: A Guide to Learning across Time and Space.* Chatham House, Chatham, New Jersey.

Rosen, A. 1962. Streets of Gold: The development and growth of the Johannesburg road system, in *IV World Meeting of the International Road Federation*, Madrid.

Roy, A. 2010. *Poverty Capital: Microfinance and the Making of Development.* Routledge, London.

Roy, A., and Ong, A. (eds.). 2011. *Worlding Cities: Asian Experiments and the Art of Being Global.* Wiley-Blackwell, Malden, Massachusetts.

Rustenburg Rapid Transit. 2013. *Taxi Operators Embark on Two-city Study Tour; Local Government Works Together to Support Public Transport Transformation.* Rustenburg Rapid Transit.

van Ryneveld, P. 2006. *Towards a Discussion on Refining the System of Decentralization and the Division of Powers and Functions between the Three Spheres of Government.* Department of Provincial and Local Government, South Africa, 1–39.

van Ryneveld, P. 2010. *Assessment of Public Transport in South African Cities*. ITDP, South Africa.

Said, A., and Herzog, A. 2011. *Lessons from How Cities Learn through Networks*. World Bank Institute Urban Division, Washington, D.C., 1–18.

Salazar Ferro, P., Behrens, R., and Wilkinson, P. 2013. Hybrid urban transport systems in developing Countries: Portents and prospects. *Research in Transportation Economics* 39, 121–132.

Samuels, M.S. 1981. An Existential Geography, in Harvey, M.E. and Holly, B.P. (eds.), *Themes in Geographic Thought*, St. Martin's Press, London.

Saunier, P.-Y. 2002. Taking up the bet on connections: A municipal contribution. *Contemporary European History* 2, 4, 507–527.

Saunier, P.-Y., and Ewen, S. (eds.). 2008. *Another Global City: Historical Explorations into the Transnational Municipal Moment, 1850–2000*. Palgrave Macmillan, Basingstoke.

Schalekamp, H., and Behrens, R. 2008. Towards a User-Oriented Approach in the Design and Planning of Public Transport Interchanges, in *Southern Africa Transport Conference*, Tshwane, South Africa.

Schalekamp, H., and Behrens, R. 2013. Engaging the Paratransit sector in Cape Town on public transport reform: Progress, process and risks. *Research in Transportation Economics* 39, 185–190.

Schalekamp, H., Behrens, R., and Wilkinson, P. 2010. Regulating Minibus Taxis: A Critical Review of Progress and a Possible Way Forward, in *Southern Africa Transport Conference*, Tshwane, South Africa.

Schalekamp, H., Wilkinson, P., and Behrens, R. 2009. *An International Review of Paratransit Regulation and Integration Experiences: Lessons for Public Transport System Rationalization and Improvement in African Cities*. African Centre of Excellence for Studies in Public and Non-Motorised Transport, University of Cape Town.

Sey, J. 2012. *The People Shall Move!: A History of Public Transport*. Department of Transport, Johannesburg, South Africa.

Shamsudin, K. 2005. Charles Compton Reade and the Introduction of Town Planning Service in British Malaya (1921–1929), in *8th International Conference of the Asia Planning Schools Association*. Kuala Lampur, Malaysia, 1–18.

Sharp, J.P., Routledge, P., Philo, C., and Paddison, R. (eds.). 2000. *Entanglements of Power: Geographies of Domination/Resistance*. Routledge, London.

Shields, G. 1979. *Port Elizabeth Tramways. A Short History Of Port Elizabeth's Road Passenger Transport Services 1879–1979*. Privately Printed, Port Elizabeth.

Shore, C., and Wright, S. 1997. Policy: A new field of Anthropology, in Shore, C. and Wright, S. (eds.), *Anthropology of Policy: Critical Perspectives on Governance and Power*. Routledge, London, 3–39.

Simmel, G. 1950 The Stranger, in translated by Wolff, K., *The Sociology of Georg Simmel*, Free Press, New York, 402–408.

Smith, J.A. 1991. *The Idea Brokers: The Rise of Think Tanks and the Rise of the Policy Elite*. Free Press, New York.

Söderström, O. 2014. *Cities in Relations: Trajectories of Urban Development in Hanoi and Ouagadougou*. Wiley-Blackwell, Malden, Massachusetts.

South African Cities Network. 2011. *State of the Cities Report*. South African Cities Network, Johannesburg, South Africa.

Spit, T., and Patton, B. 1976. *Johannesburg Tramways: A History of the Tramways of the City of Johannesburg*. The Light Railway Transport League, London.

Stanway, B. 2006. *Report on Bus Rapid Transit (BRT) Study Tour to Guayaquil and Bogota*. Department of Transport, Johannesburg.

Stanway, B. 2007. *Report on Rea Vaya BRT Visit to Bogota and Pereira in Colombia from 17th to 26th August 2007*. Department of Transport, Johannesburg.

Statistics South Africa. 2011. *Statistical release census2011*. StatsSA, Pretoria. Available at: www.statssa.gov.za October 13 2019.

Stone, D. 1996. *Capturing the Political Imagination: Think Tanks and the Policy Process*. Frank Cass, London.

Stone, D. 1999. Learning Lessons and transferring policy Across Time, Space, and Disciplines. *Politics* 19, 1, 51–59.

Stone, D. 2002. Introduction: Global knowledge and advocacy networks. *Global Networks* 2, 1, 1–12.

Stone, D. 2004. Transfer agents and Global Networks in the "Transnationalization" of Policy. *Journal of European Public Policy* 11, 3, 545–566.

Stone, D. 2010. Private philanthropy or policy transfer? The transnational norms of the open society institute. *Policy & Politics* 38, 2, 269–287.

Sutcliffe, A. 1981. *Towards the Planned City: Germany, Britain, the United States and France, 1780–1914*. Basil Blackwell Publisher, Oxford.

Tau, P. 2013. *State of the City Address*. Johannesburg, South Africa, 9 May. Available at: https://www.politicsweb.co.za/documents/state-of-joburg-address-2013-parks-tau (Accessed: 20 October 2020).

Temenos, C. 2016. Mobilizing drug policy activism: Conferences, convergence spaces and ephemeral fixtures in social movement mobilization. *Space and Polity* 20, 1, 124–141.

The Economist. 1989. Back into the Groove. *The Economist*, 26 August.

The Star. 1977. Let's Get Moving – Over to the Network. *The Star*, 29 July.

Theodore, N., and Peck, J. 1999. Welfare-to-work: National problems, local solutions? *Critical Social Policy* 19, 485–510.

Theodore, N., and Peck, J. 2001. Searching for best practice in Welfare-to-Work: The means, the method and the message. *The Policy Press* 29, 1, 81–98.

Theodore, N., and Peck, J. 2011. Framing neoliberal urbanism: Translating "common sense" urban policy across the OECD zone. *European Urban and Regional Studies* 19, 1, 20–41.

Thomas, D.P. 2013. The Gautrain Project in South Africa: A Cautionary Tale. *Journal of Contemporary African Studies* 31, 1, 77–94.

Tshwane Department of Transport. 2006. *Draft Strategy to Accelerate Public Transport Implementation via a Win-Win-Win Partnership between Government, Existing Operators and Labor*. Department of Transport, Tshwane.
Tsing, A. 2000. The global situation. *Cultural Anthropology* 15, 3, 327–360.
Turok, I., and Watson, V. 2001. Divergent development in South African cities: Strategic challenges facing Cape Town. *Urban Forum* 12, 1, 119–138.
Urry, J. 2007. *Mobilities*. Polity Press, Cambridge, UK.
Van Onselen, C. 2001. *New Babylon New Nineveh: Everyday Life on the Witwatersrand, 1886–1914*. Jonathan Ball Publishers, Johannesburg.
Venter, C. Rickert, T., Mashiri, M., and de Deus, K. 2004. Entry into High-floor Vehicles Using Wayside Platforms, in. *10th International Conference on Mobility and Transport for Elderly and Disabled People*, Japan.
Venter, C. 2013. The Lurch towards formalisation: Lessons from the implementation of BRT in Johannesburg, South Africa. *Research in Transportation Economics* 39, 114–120.
Venter, C., and Vaz, E. 2011. The Effectiveness of Bus Rapid Transit as Part of a Poverty Reduction Strategy: Some Early Impacts in Johannesburg, in *Southern Africa Transport Conference*, Tshwane, South Africa.
Vion, A. 2002. Europe from the Bottom Up: Town twinning in France during the cold war. *Contemporary European History* 11, 4, 623–640.
deWaal, L. 1973. Bus Lanes and Bus Priorities, in *Transport and Environment. South African Institution of Civil Engineers Quinquennial*, Johannesburg.
Walters, J. 2009. Overview of Public Transport Policy Developments in South Africa, in *Workshop 7: Public Transport Markets in Development. International Conference Series on Competition and Ownership in Land Passenger Transport*, Delft, The Netherlands (Thredbo 11).
Walters, J. 2013. Overview of public transport policy developments in South Africa. *Research in Transportation Economics* 39, 34–45.
Ward, K. 2006. "Policies in Motion", Urban management and state restructuring: The Trans-local expansion of business improvement districts. *International Journal of Urban and Regional Research* 30, 1, 54–75.
Ward, K. 2007a. Business improvement districts: Policy origins, mobile policies and urban livability. *Geography Compass* 1, 3, 657–672.
Ward, K. 2007b. "Creating a personality for downtown": Business improvement districts in Milwaukee. *Urban Geography* 28, 8, 781–808.
Ward, K. 2011. Policies in motion and in place: The case of Business Improvement districts, in McCann, E. and Ward, K. (eds.), *Mobile Urbanism: Cities and Policymaking in the Global Age*. Minnesota Press, Minneapolis, 71–96.
Western, J. 1985. Undoing the Colonial City? *Geographical Review* 75, 3, 335–357.
White, P.S. 2003. World's Leading Urban Reformer Brings "Bogota Model" to Africa. *ITDP Press Release*, 14, January, 1–2.
Williams, M. 2003a. Catching a Bus to the Future. *Cape Argus*, 14 April.
Williams, M. 2003b. Klipfontein Bus Plan "Poverty-Busting Driving Force". *Cape Argus*, 23 April.

Willumsen, L. (Pilo), and Lillo, E. 2003. Recommendations: Cape Town is Not Bogota. But You May Learn From Their Experience. Cape Town, 5 November.

Wolman, H. 1992. Understanding cross national policy Transfers: The case of Britain and the US. *Governance* 5, 27–45.

Wolman, H., and Page, E. 2002. Policy transfer among local governments: An information-theory approach. *Governance: An International Journal of Policy, Administration, and Institutions* 15, 4, 477–501.

Wood, A. 2014a. Learning through policy tourism: Circulating bus rapid transit from South America to South Africa. *Environment and Planning A* 46, 11, 2654–2669.

Wood, A. 2014b. Moving policy: Global and local characters circulating Bus Rapid Transit through South African Cities. *Urban Geography* 35, 8, 1238–1254.

Wood, A. 2015a. Multiple temporalities of policy circulation: Gradual, repetitive and delayed processes of BRT adoption in South African cities. *International Journal of Urban and Regional Research* 39, 3, 568–580.

Wood, A. 2015b. The politics of policy circulation: Unpacking the relationship between South African and South American cities in the adoption of Bus Rapid Transit. *Antipode* 47, 4, 1062–1079.

Wood, A. 2016. Tracing policy movements: Methods for studying learning and policy circulation. *Environment and Planning A* 48, 2, 391–406.

Wood, A. 2019a. Circulating planning ideas from the metropole to the colonies: Understanding South Africa's segregated cities through policy mobilities. *Singapore Journal of Tropical Geography* 40, 2, 257–271.

Wood, A. 2019b. Disentangling the nexus of global intermediaries: The case of bus rapid transit. *Urban Development Issues* 62, 17–27.

Wood, A. 2019c. The business of global intermediaries in the promotion of bus rapid transit, in Baker, T. and Walker, C. (eds.), *Public Policy Circulation: Arenas, Agents and Actions*, Edward Elgar Publishing, London.

Wood, A. 2020. Tracing urbanism: Methods of actually doing comparative studies in Johannesburg. *Urban Geography* 41, 2.

Wood, A., Kębłowski, W., and Tuvikene, T. 2020. Decolonial approaches to urban transport geographies: Introduction to the special issue. *Journal of Transport Geography* 88, 1–5. doi:10.1016/j.jtrangeo.2020.102811

World Bank. 2013. *Toolkit on Intelligent Transport Systems for Urban Transport: Johannesburg, South Africa*. World Bank, Washington, D.C., 1–9.

Wright, L. 2002. Bus Rapid Transit Systems in Latin America: An Overview with Special Focus on Bogota's TransMilenio System. *Centre for Transport Studies Lecture Series*, Lecture Room 10, 3rd Level Menzies Building, University of Cape Town, 9 September. Available at: http://www.ctfs.uct.ac.za/downloads/lecture/2002/wright_2002.pdf January 23 2014.

Wright, L. 2007a. *Bus Rapid Transit Planning Guide*. Institute for Transportation and Development Policy, New York.

Wright, L. 2007b. *Catalytic Public Transport Initiatives in South Africa: A Critical Review*. VivaCities, South Africa.

Wright, L. 2007c. *City of Cape Town Study Tour to Latin America: Technical Visit to Latin American BRT Systems*. Department of Transport, Cape Town.

Wright, L. 2007d. *Tshwane Rapid Transit: Implementation Framework*. City of Tshwane Metropolitan Municipality, Tshwane, South Africa.

Zhang, X., Liu, Z., and Wang, H. 2014. Lessons of Bus Rapid Transit from Nine Cities in China. *Journal of the Transportation Research Board* 2394, 45–54.

Zheng, L., and Jiaqing, W. 2007. Summary of the application effect of Bus Rapid Transit at Beijing South-Centre Corridor of China. *Journal of Transportation Systems Engineering and Information Technology* 7, 4, 137–142.

Index

NB Page locators in *italic* denote illustrations.

Accra, Ghana, 103, 104
African National Congress, 109
Agencement en Rames Automatisées de Modules Indépendants dans les Stations (Aramis), Paris, 135
Ahmedabad, India, 53, 105
Alexandra bus boycotts, 4
apartheid
   pass system, 3
   racial categories, 15n3
   resistance to, 59, 81
   segregation, 59, 121
   South African exceptionalism, 102
   transport legacy, 2–4, 59–60, 120–128
   urban planning, 2–3, 6, 42
A Re Yeng transport system, Tshwane, 41–44, *42, 43*
   construction, 43–44, 155
   objectives, 44
   operating companies, 159
   phases, 157
   problems, 43
automobile reliance, 61, 69n4, 145
   American model, 87
   and city form, 3, 4
   and social aspiration, 4

Barcelona, 10
Behrens, Roger, 129, 130, 131
Bennett, Colin J., 5
Bethlehem, Lael, 52, 53, 58
Bilbao, 19
Bing Thom architects, 20–21, 23
Black entrepreneurialism, 59
   informal taxi industry, 4, 37–38, 59
Bogotá, Colombia, 1, 30
   BRT model *see* Bogotá model
   bus segregation, 32
   policy tourism *see* Bogotá tours
   rebranding, 99, 102
   Transmilenio *see* Transmilenio transport system, Bogotá
   transport planning, 32–33, 101
Bogotá model, 1–2, 27, 30, 119, 138
   and African BRT, 2, 20, 52, 101, 102, 130, 133
   formation, 31
   representational power of, 1, 102

variations, 31, 34, 49, 102, 103
  see also Bogotá tours
Bogotá tours, 2, 88, 89, *91*, 93–94
Boraine, Andrew, 78–79, 82, 94
Browning, Paul, 76–77
BRT adoption, in South Africa, *32*, 36, 119, 124, *124*, 126
  delays, 130–133
  grants for *see* Public Transport Infrastructure and Systems Grant (PTISG)
  international influences *see* policy tourism
  management, 68–69
  pace of, 119, 120, 125, 141
  phases, 114–115
  variations, 48, 143, 115
BRT in African cities, 103–104
  Dar es Salaam, 103, 104–105, *104*
  Lagos, 100, 103–104, 107
  see also South African BRT systems
BRT lite, 100, 103–104
  differences from full BRT, 29, 103
  rejection by South African cities, 103, 105, 107
BRT models, 28, 31, 68
  African, 103–105
  Bogotá *see* Bogotá model
  BRT lite *see* BRT lite
  South American, 98–102
BRT stations, 44, 51, 162–163
  costs, 37, 38, 44, 49
  design, 51–52
  illustrations of, *38, 40, 43, 45*
  locations, 38, 40, 43
  platforms *see* station platforms
business improvement districts, 6
bus lanes, 53–56, 127–128, *127*, 160–161
  Kassel curbs, 54–55, 160
bus rapid transportation (BRT) systems, 1–2, 28, 50, 57–58, *66*
  affordability, 138
  in Africa *see* BRT in African cities
  concept, 1, 29–30, 50
  funding for, 30, 32–33
  global prevalence, 1–2, 27, 28–29
  government interference, 43
  in India, 53, 88, 105–106
  infrastructure, 31, 33, 130
  lane segregation, 53–54, *127*
  in Latin America, 28, 88–93
  local considerations, 58, 74
  management, 65, 68, 146
  models *see* BRT models
  mutability, 27
  operators, 65, 67–68, 91
  as planning opportunity, 65
  popularity, 1, 4–5, 27, 50, 138
  ridership figures, 30, 31, 52–53, 145, 164–165
  routes, 55–58
  in South Africa *see* South African BRT systems
  standing passengers, 52, 162
  stations *see* BRT stations
  variations, 31, 68
  vehicles, 52–53, *53*
bus transportation, 3, 4
  BRT *see* bus rapid transportation (BRT) systems
  diesel, 123
  minibuses *see* minibus taxis
  tramways *see* tramways
  trolleybus, 3, 30, 102, 123
  see also public transportation

Cameron, Bill, 133
Cape Town, 3, 39
  adoption of BRT, 39–40, 58, 93, 114
  African townships, 39
  Bogotá tour, 89, 90
  city improvement districts, 6
  demography, 39
  transport systems *see* Cape Town public transportation
Cape Town public transportation, 3, 36, 39, 48, 60, 146
  BRT adoption, 2, 131, 145
    see also MyCiTi transport system, Cape Town
  bus lanes, 128
  integrated systems, 48
  intelligent transport system (ITS), 41, 95

INDEX   187

Klipfontein Road *see* Klipfontein
   Corridor project, Cape Town
   management, 65, 116, 132, 146
   policy, 130, 132
   rail, 60, 61, 116
   taxi operators, 65
   tramways, 122
Caracas, 24
Centre for Transport Studies (ACET),
   Cape Town, 129, 131
Chicago, USA, 29
Chinnappen, Eddie, 55, 84–85, 101, 110
Chua, Beng-Huat, 19
cities, South African, 6
   BRT development *see* South African
      BRT systems
   collaboration, 110, 112
   colonial legacy, 6
   competition between, 107–108, 111
   dualisms, 3
   and foreign planning models, 6, 100
   historical background, 2–3
   mobility dynamics, 3
   political mindset, 34
   see also South African Cities Network
      (SACN); *and individual cities*
Colombia, 126
   Pereira *see also* Pereira, Colombia
colonialism, 6, 107
   and north–south imbalance, 24
   planning practices, 72, 102–103
Commonwealth Games projects, 21
Costanza, Madeleine, 76, 129,
   131, 136
Cronin, Jeremy, 47–48, 79, 83,
   100, 137
   and ITDP, 87–88
   and study tours, 89
Curitiba, Brazil, 33, 129, 133
   BRT system *see* Rede Integrada de
      Transporte, Curitiba
   and Lima's transport system, 101
   study tours, 29, 33, 88–89, 90, 124,
      125, 129–130, 144
   World Bank Case Study, 129
cycleways, 32, 34, 75, 146

Dakar, Senegal, 103, 104
Dar es Salaam, Tanzania, 103,
   104–105, *104*
de Waal, Louis, 127
Drost, Bea, 86
Durban *see* eThekwini

Eastern Cape province, 45
Ekurhuleni municipality, Gauteng, 36,
   110, 111, 113
Embarq, 32, 86, 89, 92
Essop, Tasneem, 131, 132
eThekwini, 2, 36, 46–47, 100
   adoption of BRT, 47, 58, 115–116
   Go Durban! *see* Go Durban! transport
      system, eThekwini
   rail, 64, 108
   study tour, 90
ethnography, multi-sited, 9

fast policymaking, 25–26, 138
   literature on, 14, 25, 26, 138
   term, 25, 125
Fortune, Gershwin, 136
Fort Worth, Texas, 23
Foucauldian theories, 21–22
Frieselaar, Andre, 113, 128, 137
Froschauer, Pauline, 81, 119
   BRT transport manager, 45, 81,
      94–95
   on funding, 111–112, 137
   as intermediary, 81
   on rate of BRT adoption, 125

garden city planning model, 121
Gauteng Freeway Improvement Project
   (GFIP), Johannesburg, 37
Gauteng province, 36, 110, 111,
   116–117, 136
   household travel survey, 145
Gautrain project, 134, 136–137
   financial viability, 137–138
global north
   power imbalance with the south, 24
   and South Africa, 106–107
   see also colonialism

global south
  city planning, 6, 24
  regional leadership, 106
  transportation information distribution, 7
Go Durban! transport system, eThekwini, 47–48, 155, 157
  construction, 155
  management, 157
  modal split, 157
  operating companies, 159
  services operating, 157
González, Sara, 19
Gordge, Richard, 108
Gotz, Graeme, 50
Guangzhou, China, 30
Guayaquil, Ecuador, 30, 92

Haiden, Ron, 12–13, 83, 84, 123, 125
Harrison, Philip, 103–104, 105–106, 129
Herron, Brett, 109, 110, 115
Hoffman, Lisa, 19
Hook, Walter, 79, 83, 87–88, 103

India, 53, 88
  links with South Africa, 105–106
informal transportation, 59–60, 139, 143
  and BRT, 13, 28, 51, 59, 65–66
  networks, 92
  quality, 4, 59
  see also minibus taxis
information exchange, between cities, 5–6
  in Africa, 103
  described by Herodotus, 5
  as municipal diplomacy, 23
  in Palmyra, 5
  poverty alleviation, 23
  south–south, 6–7, 8, 24, 98–99, 100
  in St. Petersburg, 5
  see also information networks
information networks, 72, 85–96, *86*
  classic analysis *see* social network analysis
  formation, 85
  informal, 71
  international, 86–88
  local, 88–92
  power dynamics, 94–95

innovation in public transportation, 33, 120–139
  BRT awareness, 124, *124*, 126
  colonial legacy, 6, 121, 123
  competitive, 117
  and development projects *see* transit-oriented development
  fads, 133–134
  international influences, 9, 19, 72, 103, 125
  as urban solution, 134
Institute for Transportation and Development Policy (ITDP), New York, 1, 31, 87–88
  BRT promotion, 33–34, 87
  BRT standards, 48
  criticism of, 88
  formation, 87
  study tours, 92, 100
intelligent transport systems (ITS), 1, 94
  Cape Town, 41, 95
international BRT consultants, 1, 71, 72, 74, 75–78, 94
  see also individual consultants
international transport organizations, 32
  promotion of BRT, 31–32
interviews, as methodology, 10, 11–12, 15n2, 147–153
Istanbul, Turkey, 30

Jakarta, Indonesia, 30
Janmarg BRT system, Ahmedabad, 105
Johannesburg, 3, 13, 36–37
  and Bogotá tour, 89, 90
  bus boycotts, 4
  city development strategy, 6, 7
  Corridors of Freedom, 146
  demographics, 36, 37
  economic disparities, 36–37
  transport systems *see* Johannesburg public transportation
Johannesburg public transportation, 2, 3, 13, 60
  BRT adoption, 37–39, 58, 92, 114, 130–131, 145

see also Rea Veya transport system, Johannesburg
   bus lanes, 127
   and development projects, 146
   Metrobus, 38, 65, 66, 67, 158
   monorail project, 126, 135–136
   rail, 60, 61
   tramways, 121, 122
Johnson, Boris, 24
Jones, John, 84, 126

Kassel curbs, 54–55
Khosa, Meshack M., 59
Klipfontein Corridor project, Cape Town, 64, 124, 126, 131–132
   failure, 132
   feasibility study, 77
   inclusion in MyCiTi routes, 64
   influence of, 132
Klipfontein Road *see* Klipfontein Corridor project, Cape Town
Krogscheepers, Christoph, 95
Kumar, Ajay, 50, 88, 110
KwaZulu-Natal province, 46

Lagos, Nigeria, 100, 103–104
Larner, Wendy, 20, 21, 22
Latour, Bruno, 135
Laurie, Nina, 20, 21, 22
Law, John, 9
Libhongolethu transport system, Nelson Mandela Bay, 45–46
   bus lanes, 55
   construction, 46
   costs, 46
   management, 157
   modal split, 157
   objectives, 46
   operating companies, 159
   phases, 155
   problems, 46
   routes, 157
   and World Cup 2010, 64
Lima, Peru, 29, 101
Litman, Todd, 74, 82
Litsamaiso transport operating company, 38, 65, 158

London, UK, 24, 26
Los Angeles, USA, 29

Macdonald, Laura, 23
Mahon, Rianne, 23
Manchester, UK, 21
Manning, Jonathan, 111
Manyathi, Thami, 65, 115–116
Manzana, Nkosinathi, 109–110
market liberalization, 60
Masondo, Amos, 51, 109
Massey, Doreen, 24
materiality, of projects, 28, 68, 70
   BRT, 55, 68
   and circulation, 28
   socio-materiality, 12
Mazaza, Maddie, 50–51, 110
Mbembe, Achille, 7
Mbete, Baleka, 109
McCann, Eugene, 10, 20–21, 71
   policy boosterism, 21
McCaul, Colleen, 112, 128–129
McFarlane, Colin, 22
Metrobus, Johannesburg, 38, 65, 66, 67, 158
Metrorail system, 56, 57, 116
Metshwane, Eric, 84, 101, 102
Mexico City, 23, 53
minibus taxis, 3, 4, 38, 58, 59–61
   competition for passengers, 59, 60
   inclusion in BRT systems, 38, 41, 45, 59–61, 65, 66–67, 93
   operator organizations, 38, 41, 45, 65, 69n6
   popularity, 60
   state regulation, 59, 62–63
   system analysis, 77
mobilities research methodology, 11–13
   analytics, 71–73, 75
   interviews, 10, 11–12, 15n2, 147–153
   tracing *see* tracing methodology
modernism, 4, 6, 7
Moodley, Logan, 100
Moosajee, Rehana, 52, 54, 66, 82
   attack on, 142
   and Brett Herron, 109, 110
   on policy exchange, 95, 125

on policy mobilizers, 76
study tour, 92
support for BRT, 50, 82, 114
and taxi industry, 82
on World Cup hosting, 64–65
Motor Carrier Transportation Act (1930), 59
municipalities, South African, 45–46, 69n3
    grants to, 63
    integrated transport plans (ITPs), 62
    see also individual municipalities
MyCiTi transport system, Cape Town, 12, 13, 39–41, *40*, 156
    bus lanes, 55, 160
    construction, 40–41
    costs, 40–41, 78
    features, 49
    local partnerships, 93–94
    management, 156
    modal split, 156
    objectives, 40
    operating companies, 158
    planning stages, 132
    problems, 41
    routes, 55–57, 56, 64
    services operating, 156
    taxi operators, 65
    vehicles, 52–53

Naidoo, Subethri, 133–134
Nairobi, 103, 104
Namela transport consultants, 45, 81
National Department of Transport, South Africa, 61–62, 64, 69n6, 113
National Land Transport Transition Act (200), 62, 116
Nelson Mandela Bay, South Africa, 2, 36, 45–46
    study tour, 90
    transport systems, 46, 56
networks
    information *see* information networks
    policy, 9, 70, 72, 77
    transport *see* public transportation
Neumann, Sigismund, 121

New York City, 53
New Zealand urban policy, 20, 21
Nkhahle, Seana, 95
North West Province, 44

Olympic Games projects, 21
Ong, Aihwa, 34

Peck, Jamie, 9–10, 133
Penalosa, Enrique, 1, 12, 31, 32, 75
    in India, 105
    and ITDP, 12, 31, 33–34, 87
    outside perceptions of, 76, 77
Penalosa, Gil, 12
Pereira, Colombia, 133
    opposition to BRT, 92
    study tours, 92, 101, 102
    taxi operators, 90, 92, 93, 101–102, 133
personal rapid transit (PRT) technologies, 134–135
Peter the Great of Russia, 5
Pieterse, Edgar, 85–86
PioTrans, 37–38, 65, 67, 84
Pneuways Company, Zimbabwe, 134
policy actors, in BRT projects, 20, 22, 68–69, 70
    decision makers, 20, 22
    definition, 71
    intermediaries, 76–81
    international consultants, 1, 71, 72, 74, 75–78, 94
    local pioneers, 81–85
    mobilizers, 22–23, 70, 72, 75–78
    and networks *see* information networks
    roles, 72, 74–75, 142
    types, 71–72, 73, 75
policymaking, 25–26, 71, 143
    actors *see* policy actors, in BRT projects
    crisis-driven, 25–26
    see also fast policymaking
    incremental, 25, 101
policy mobilities approaches, 7–8, 9–13, 16–17, 18, 70
    follow the policy, 9–10

follow the project, 9, 10
historical, 119–120, 138
metaphors, 24–25
multidisciplinary, 17–18
research methodology *see* mobilities research methodology
policy mobility, 18
  actors *see* policy actors
  and information exchange *see* information exchange, between cities
  and local politics, 97
  scholarly approaches *see* policy mobilities approaches
  temporalities *see* temporalities, in policy adoption
policy models, 30–31
  agency of, 20, 68
  BRT *see* BRT models
  etymology, 18
  as mobile objects, 19–20
  see also policy mobility
  modifications, 20, 34
  origins, 19
policy sharing, 24–25
  see also information sharing, between cities; policy mobility
policy tourism, 88–93
  criticism of, 89
  study tours *see* study tours
political cartoons, 65
politicians, South African, 76, 83
  as BRT managers, 68
  as BRT policy actors, 27, 81–82
  competition between, 108, 109
  information gathering, 99, 113
  on South American study tours, 2, 88, 89, 101
  see also individual politicians
Port Elizabeth *see* Nelson Mandela Bay
post-apartheid movements, 5, 81
poverty alleviation, 23
power relations, 21–22
  and BRT adoption, 94, 96, 117
  global system, 23
  institutional, 22
  and knowledge, 21–22

Pretoria *see* Tshwane
Pretorius, Lynne, 132, 133
provinces, South African, 39, 116
  governments, 62
  responsibility for transport, 61, 80, 136
  rivalry, 107, 109
  see also individual provinces
Public Transport Action Plan (2007), 62–63, 81
public transportation, 61–65
  BRT systems *see* bus rapid transportation (BRT) systems
  Cape Town *see* Cape Town public transportation
  consumer uptake, 61
  decentralization, 62
  high cost, 61
  history of *see* innovation in public transportation
  Johannesburg *see* Johannesburg public transportation
  problems, 52, 64
  responsibility for, 61–62
  role of entrepreneurs, 123
  state intervention, 61–65
  subsidies, 3, 64, 107
  types *see* transportation, modes of
  and World Cup, 37, 46, 57, 64, 133
Public Transport Infrastructure and Systems Grant (PTISG), 63–64
  creation of, 83
  funding policies, 111–112
  grant allocation, 63, 108
  purposes, 63, 133
  and study tours, 90

Quito, Ecuador, 30, 33, 53, 102, 133
  El Trole, 102

Rabinovitch, Jonas, 129
rail transport, 4, 47, 61, 134
  commuter, 100
  high-speed *see* Gautrain
  metro system *see* Metrorail
  monorails, 134, 135

192 INDEX

networks, 4
relative cost, 61
reliability, 65
ridership, 64
structure, 4
subsidies, 61, 64, 65
technology, 133
Rea Veya transport system, Johannesburg, 37–39, *38*, 48, 156
  antecedents, 127
  architecture, 38
  bus lanes, 53, 127, 160
  construction, 37, 38–39
  features, 49, *52*
  influence on other BRT systems, 113–114
  interviews with bus operators, 12
  management, 156
  modal split, 156
  operating companies, 38, 65, 67–68, 158
  passenger comfort, 52
  routes, 57
  services operating, 156
  similarity to Jambarg, Ahmedabad, 105
  vehicles, 52, *53*
  World Cup availability, 37, 57, 64
Rede Integrada de Transporte, Curituba, 29, 30, 129
  influence on South African BRT, 29, 33, 51, 120, 125
Road Transportation Bill (1977), 59
Robinson, Jennifer, 17, 19, 24
Rose, Richard, 5
Rustenburg, 2, 36, 44
  Bogotá tour, 89, 90
  and Froschauer, 45, 81, 94–95
  relations with other cities, 113
  transport systems, 44–45, 55, 81

Sao Paolo, Brazil, 31
Seedat, Ibrahim, 39–40, 76, 80, 116, 137
Seedat, Rashid, 58, 101
Sexwale, Tokyo, 136
Shilowa, Mbhazima, 136
Siemens, 134
Simmel, Georg, 12

Singapore model, the, 18–19, 26
social network analysis, 85
  political mindset, 34
South African BRT systems, 35, 35, 36, 27–69
  adoption *see* BRT adoption, in South Africa
  bus ownership, 160–161
  costs, 40–41, 46, 78, 164–165
  features, 48, 154–165
  financial viability, 145–146
  history *see* innovation in public transportation
  influence of Transmilenio *see* Bogotá model
  and international consultants, 1, 71, 72, 74, 75–78, 94
  local politics of, 98, 107–114
  maps, *35*, *56*, *57*
  number of buses, 160–161
  physical features, 58
  popularity with residents, 50
  popularity with urban planners, 4–5, 27
  relations with taxi industry, 59–61, 93, 113, 142
  routes, 56–58
  variations, 48, 143
  see also individual BRT systems
South African Cities Network (SACN), 79, 112–113
South African National Taxi Council (SANTACO), 66, 69n6
South African political parties, 109
South African Transport Conference (SATC), 2
  2006 workshops, 36, 74, 140
  2012 workshops, 113
Spencer Commission, 123
Stanway, Bob, 83–84, 92, 95, 101
station platforms, 30, 51–52
  high-floor, 49, 51, *52*, 162–163
  low-floor, 43, 44, *45*, 47, 51, 162–163
  see also BRT stations
St. Petersburg, Russia, 5
Strategic Public Transport Network (SPTN), 130–131
study tours, 89–92, 129

Bogotá, 2, 88, 89, *91*, 93–94
Curitiba, 29, 33, 88–89, 90, 124, 125, 129–130, 144
ITDP, 92, 100
Pereira, 92, 101, 102
sustainability, 2, 19, 75
  Singapore model, 19
  transportation, 33, 82, 86–87, 116
  Vancouver model, 18
Sutcliffe, Michael, 109

Tau, Parks, 146
Taxi Recapitalization Program (TRP), 62, 67
taxis
  associations, 38, 41, 45, 65, 69n6, 92
  incorporation into BRT, 65–67
  pre-BRT system, 59–62, 77
  state regulation, 62, 65
  see also minibus taxis
temporalities, in policy adoption, 26, 119–139
  rates of change, 120
Theodore, Nik, 9–10
Thom, Bing Wing *see* Bing Thom architects
Thompson, Elizabeth, 109
Tofie, Zaida, 132
Toronto, Canada, 23
tracing methodology, 16–26
  through actors and associations, 20–23
  through cities, 23–25
  through policy models, 18–20
  through temporalities, 25–26, 120
tramways, 3, 121, 122, 133
  electric, 122–123, *122*
  horse-drawn, *121*, 122
transit-oriented development, 133, 135–136, 146
Transmilenio transport system, Bogotá, 1, 27, 31, 33
  business model, 93
  bus lanes, 53
  as inspiration for BRT systems *see* Bogotá model
  objectives, 31, 33
  platforms, 51

problems, 33
routes, 57
transportation, modes of, 3–4, 60, 61, 121–122, 134
  cycling, 32, 34, 61, 75, 146
  minitrams, 134
  monorails, 134, 135
  taxis *see* taxis
  tramcars, 3, 121, *121*, 122
  trolleybuses, 3, 30, 102, 123
  see also automobile reliance; public transport; rail networks
transportation best practice, 1–2, 9, 101, 139
  adoption, 13–14
  circulation, 21, 24, 26, 70, 77, 88, 142–143
  and information exchange *see* information exchange, between cities
  localization, 8, 14, 72, 88, 107, 121, 138
  policy models, 96
  quick fix, 119, 125, 138
  tailored, 78
  templates, 20
  theoretical, 13
transportation planning, 1, 4, 60
  best practice *see* transportation best practice
transport geography, 7–8
  decolonial approach, 7
  information exchange *see* information change, between cities
  mathematical modeling, 7
  of South African BRT, 27–69
Transport and Urban Development Authority (TDA), 146
trolleybuses, 3, 30, 102, 123
Tshwane, 2, 36, 41, 90
  adoption of BRT, 42, 115, 145
  leadership, 115
  transport systems, 41–44, *42*, 114
  see also A Re Yeng transport system, Tshwane
Tsing, Anna, 24
urban planning, 1, 6, 8, 146

information sharing *see* information exchange, between cities
policymaking, 21
transportation *see* transportation planning
Urry, John, 9

Vancouver model of sustainability, 18
Vancouverization, 21, 23
van Ryneveld, Philip, 79, 116
  and Helen Zille, 79, 80
  introduction of BRT, 11, 79–81
Volvo Research and Education Foundation (VREF), 32, 86
Vorster, Hilton, 116

Walters, Jackie, 129–130
Ward, Kevin, 10, 21
Western Cape Province, 39, 80, 109, 132
White, Paul Steely, 75–76, 87
Williumsen, Pilo, 77
World Cup (football)2010
  and BRT promotion, 125
  South African hosting, 112, 132–133, 137
  transport for, 37, 46, 57, 64, 133

WRI Centre for Sustainable Transport *see* Embarq
Wright, Lloyd, 74, 75, 76, 95
  BRT Planning Guide, 76
  BRT promotion, 2, 39–40, 75, 76, 80, 131
  at MyCiTi, 78
  on pace of BRT adoption, 126
  as policy mobilizer, 11, 12, 74, 76–77, 78, 130
  on South African politicians, 82
  study tour leader, 92
  and Tshwane Rapid Transit Framework, 42–43
  and VivaCities, 76

Yarona transport system, Rustenburg, 44–45, 155
  construction, 45, 155
  operating companies, 159
  phases, 157, 159, 161, 163

Zille, Helen, 40, 79, 80, 82, 109, 115
Zimmerman, Sam, 138
Zuma, Jacob, 142